National Aeronautics and Space Administration

Fairing Well

Aerodynamic Truck Research at NASA's Dryden Flight Research Center

Christian Gelzer

National Aeronautics and Space Administration
NASA History Office
Washington, D.C.
2011

Table of Contents

Acknowledgments v

Introduction vi

Chapter One 1
Drag

Chapter Two 3
Eddies and Currents

Chapter Three 11
The Shoebox

Chapter Four 17
Word Spreads

Chapter Five 25
Shifting the Paradigm

Chapter Six 37
Technology Transfer

Chapter Seven 43
Depressed Cows

Chapter Eight 49
Laws Change; Physics Doesn't

Chapter Nine 51
The Drag Bucket

Chapter Ten 55
Results

Chapter Eleven 67
The Social Construction of a Technology

Appendices 71

    Appendix A, aerodynamic concepts from patents for tractor trailer vehicles 72

Appendix B, U.S. Patent 4,343,506 125

Appendix C, U.S. Patent 6,892,989 132

Bibliography 139

Index 148

The NASA History Series 151

Acknowledgments

Completing this manuscript meant the participation of many. I am indebted to Tom Tschida of the Dryden Photo Lab for the more recent pictures and their scans, as well as Tony Landis for his work on the scans that appear in this book; their work never disappoints. Jay Levine laid out the manuscript with his usual panache. He and others helped with the title selection, which posed more than a trifling dilemma. Curtis Peebles and Peter Merlin read portions of the manuscript and provided useful comments, as well as details that bolster the story. And Sarah Merlin edited the book. Every publication from Dryden to which she applies her pen emerges vastly improved, and readable to boot, and this is no exception.

My thanks also extend to my peer reviewers who carved time from their schedules both to read the manuscript with care and to suggest critical changes that greatly enhanced the final product. They kept me honest when it came to numbers. Time spent with them was a privilege.

And I am obliged to those from the trucking industry—manufacturers, collectors, and end users—who freely offered up ideas and comments when I asked; this would have been an incomplete work without them.

But I am especially indebted to Ed Saltzman, who regularly took time to look for documents, to explain the arcane, and to correct my errors as the manuscript developed. If there is a genesis to this project, it is Ed.

Christian Gelzer
NASA Dryden
Edwards, California

Introduction

It is a central tenet among historians of the subject that technology does not exist or develop in a vacuum: it is advanced, inhibited, or redirected by social influences. This approach to the subject is referred to as the "social construction of technology." Melvin Kranzberg, an early, pivotal figure in this field of history, proposed six laws on the subject that expound on the opening statement. Among the laws he postulated: technology was neither good nor bad, nor neutral; technological change is not inevitable; and invention is the mother of necessity.[1] Kranzberg laid out his laws partly tongue-in-cheek, but he was quite serious about their underlying rationale, namely, that as a human creation, technology was subject to human vicissitudes, undermining the notion that technology had its own inherent logic that determined the course of development once introduced. The idea of technology with an inherent logic defying human involvement is referred to as "technological determinism."[2] Over the years historians have produced a mounting collection of scholarly works informed, directly or indirectly, by the argument for the social construction of technology.

Donald MacKenzie, for example, notes in his *Historical Sociology of Nuclear Missile Guidance* that it is wrong "to assume that missile accuracy is a natural or inevitable consequence of technical change. Rather, it is the product of a complex process of conflict and collaboration between a range of social actors."[3] There was no predicating assumption that nuclear-tipped ballistic missiles needed accuracy. The creation of an inertial guidance system for ballistic missiles postdated the missiles' creation, and not everyone in a decision-making position in the U.S. military's nuclear weapons program was convinced of the need for such an invention. Guidance systems are now an essential part of these delivery devices nonetheless. Coleen Dunlavy offers another example of this in her examination of nineteenth-century railroads on both sides of the Atlantic Ocean. Her work reveals the extent to which cultural influences determined how railroads developed instead of any logic embedded in the technology. Or consider Thomas P. Hughes, who found that when it came to urban electrification there was no logical pattern for establishing electrical grids intrinsic to the technology; instead, individuals and cultures were key factors in determining how such grids would evolve, and which form of electricity (AC or DC) would dominate.[4]

One of the common assumptions that historians such as Kranzberg, Hughes, and so many others in the discipline struggled against is the idea that technological development moves in a mostly linear progression, typified by A to B, followed by C, D, then E. This notion is, of course, part and parcel of technological determinism, and something that social construction vitiates. Rather, scholars have shown that technology's development often bears an erratic pattern, somewhat like A to C, then E, then B, and maybe D, a pattern that defies rational development.[5]

The American civilian space program and the "race

---

[1] Melvin Kranzberg, "Presidential Address," *Technology and Culture,* vol. 27, no. 3 (July 1986), (Baltimore: The Johns Hopkins University Press, 1986), pp. 544-560. The notion that invention is the mother of necessity was meant to be provocative. Study has shown it to be surprisingly common.

[2] Technological determinism asserts there are inevitable results once a technology is introduced. In such an interpretation, it is technology that directs the path of human activity and shapes society. Few scholars champion technological determinism in its purest form, but there are those who give primacy to technology's role in social change, which is the inverse of today's dominant school of thought. A classic example of this is Lynn White's *Medieval Technology and Social Change.* In this short but remarkably dense book White traces the evolution of the stirrup from Asia to Europe, and then the stirrup's role in the rise of feudalism in European society. White was by no means a strict determinist, but the core of his argument is that the stirrup's arrival in Europe changed warfare, which in turn led to the reordering of society to accommodate the demands this new style of fighting required. See Lynn White, Jr., *Medieval Technology and Social Change* (London: Oxford University Press, 1962).

[3] Donald MacKenzie, *Inventing Accuracy: A Historical Sociology of Nuclear Missile Guidance* (Cambridge: MIT Press, 1993), 3.

[4] Thomas P. Hughes, *Networks of Power: Electrification in Western Society, 1880-1930,* 2nd ed. (Baltimore: Johns Hopkins University Press, 1983). See also Hugh G. J. Aitken, *Scientific Management in Action: Taylorism at Watertown Arsenal, 1908-1915* (Princeton: Princeton University Press, 1960). Aitken showed the influence a group of skilled workers had over the imposition of F. W. Taylor's process management system, a process that was often seen as so compelling in logic that its incorporation was insurmountable.

to the moon" are fine examples of this. The American plan consistently outlined by early advocates was: develop a rocket capable of achieving orbital velocity, then develop the ability to carry equipment into Earth orbit where a space station could be assembled, after which exploratory missions could then be launched into our solar system from that station. The plan's logic was compelling. An orbiting station would allow for differently sized rockets to be assembled for different missions while material and fuel could be positioned there instead of launched *in toto* from Earth for each mission.[6]

This plan was turned on its head suddenly, in 1961, when president John F. Kennedy took a nascent U.S. space program and made it into a political tool. He and his advisors realized that the early Soviet successes in space were a potent tool in the Cold War, and had to be met in kind. As a result, what had once been a logically planned technological undertaking that everyone expected would proceed from A to B to C to D and then E suddenly went from A to B and then E, skipping C and D altogether. Now there would be no space station and no pre-positioning of equipment for exploration; instead, NASA would go from Earth orbit to the moon with nary a stop along the way.[7]

Eventually the agency would reach step C—the space station—well after the other goals were behind it, but by then much of the American public would wonder about the value of going back to fulfill that objective.

Cultural influences can be obvious and apparent or they can be subtle and nearly invisible, but they are always present. Take Robert Moses, for example, a noted urban planner in the first half of the twentieth century who worked in New York City and its environs. Moses was an early advocate of the automobile and by the end of his career was responsible for many of the elevated roadways that bisect the city's five boroughs, arteries that cross whole sections of the city, giving drivers the ability to traverse the metropolis without stopping in it.

Early in his career, in his position as president of the Long Island Park Commission, he saw to the construction of new roads from New York City to parks and beaches outside the city so that residents could find relaxation in the open air. His work in this instance placed Moses in opposition to wealthy landowners on Long Island who begrudged a loss of exclusivity; meanwhile, touting the automobile and access to these destinations made him the champion of the Everyman. Or so it seemed. Unnoticed by virtually everyone was that Moses made sure that the bridges crossing the New York State Parkway (one of the early modern limited-access roads), the same road delivering New Yorkers to these parks and beaches, were high enough for cars to pass under, but not buses. This ensured that that only the "right sort" would reach the parks and beaches, by which Moses meant those who did not need to ride a bus to get someplace, an economic

---

[5] In a presentation on the history of the shuttle, former NASA Deputy Administrator Dale D. Meyers was very clear about the social influences that determined the development of the shuttle program and even the shuttle itself. See Jeffrey Hoffman, *16.885J Aircraft Systems Engineering, Fall 2005* (Massachusetts Institute of Technology: MIT OpenCourseWare), http://ocw.mit.edu (accessed 26 May 2011). License: Creative Commons BY-NC-SA. Lecture 1: "The Origins of the Space Shuttle," by Dale D. Meyers.

[6] In 1952 *Collier's* magazine published a six-part article on space travel. Extensively illustrated and done with Wernher von Braun, Willy Ley, and artist Chesley Bonestell, the series, which ran over a two-year period, sparked interest and enthusiasm throughout the American public. The cover of the first in the series had a winged rocket roaring over Earth with this teaser in the upper right-hand corner: "Man Will Conquer Space Soon: Top Scientists Tell How in Startling Pages." An illustration accompanying von Braun's article shows a two-ringed space station in orbit. The April 30, 1954, issue of the magazine even asked: "Can We Get to Mars?" Walt Disney capitalized on this growing enthusiasm with *Walt Disney's Man in Space* (1955), a series of films that captured von Braun, Ley, and Heintz Haber outlining their plans and concepts as well as the hurdles of space travel and exploration, all supplemented with models of spacecraft. The second in the series, "Man and the Moon," announced a clear objective and progression beyond Earth orbit; "Mars and Beyond" was the third installment. The films were broadcast on television and included animation of a multi-stage rocket, living habitats, a space suit, and illustrations of the physics involved in such a project. An estimated 42 million people saw the films. Together, these two productions are credited not merely with introducing the American public to the idea of space travel but with cementing both an enthusiasm for the idea and clear notions of how the undertaking should proceed. Wernher von Braun, "Crossing the Last Frontier," *Collier's*, March 22, 1952, *Collier's* April 30, 1954, *Walt Disney's Man in Space* (1955), and Mike Wright, "The Disney-Von Braun Collaboration and Its Influence on Space Exploration," http://history.msfc.nasa.gov/vonbraun/disney_article.html (accessed 28 April 2011).

[7] To be sure, NASA, which President Dwight D. Eisenhower created in response to the Soviet Union's launch of Sputnik 1 and II, was itself political inasmuch as he insisted the new agency rely on Vanguard, a booster not directly associated with the threat of intercontinental nuclear war.

and racial filtering. The point here is twofold: first, technology (a bridge, in this case) is not necessarily straightforward and apolitical but rather is adapted, used, modified, or even rejected by social forces; and second, those forces can be hard to spot.[8]

The aerodynamic efficiency of long-haul trucks may seem to be only a matter of fuel prices, drag coefficients, and vehicle modifications, but a nuanced examination of the subject reveals cultural influences on a technological development that might otherwise appear logical and straightforward.[9]

There are two principal objectives to this monograph. The first is to bring long overdue attention to research done at NASA Dryden on truck aerodynamics, work not usually associated with the agency but results from which had and continue to have a direct benefit to the U.S. economy. The second is to use this case as on opportunity to tease apart some of the strands of the social fabric in technology's construction and adaptation, something not regularly done with NASA's technical work. If we genuinely seek to understand ourselves, we can ill afford superficial attention to technological choices we—or others—make regarding its use, rejection, adoption, or adaptation. It's important to know who makes what decisions: the results can be surprising.

---

[8] "[Moses] instructed Shapiro to build bridges across his new parkway low—too low for busses to pass under." Robert A. Caro, *The Power Broker: Robert Moses and the Fall of New York* (New York: Vintage Books, 1975), 312.

[9] For more on this subject see Langdon Winner, *The Whale and the Reactor: A Search for Limits in an Age of High Technology* (Chicago: the University of Chicago Press, 1988), Donald McKenzie and Judith Wajcman, *The Social Shaping of Technology* (Milton Keynes, England: Open University Press, 1985), and Ruth Schwartz Cowen, *More Work for Mother: The Ironies of Household Technologies, from the Open Hearth to the Microwave* (New York: Basic Books, 1983) as a sample.

Chapter One
Drag

NASA has long been involved in projects with only tangential links to aeronautics or space. Often this is the result of serendipitous discoveries, ideas and technology for which engineers find uses other than what was first intended.[1] Such is this story.

From 1946 to 1958, the year in which the agency became NASA, the National Advisory Committee for Aeronautics (NACA) operated the High Speed Flight Research Station at Muroc Air Force Base (today's Edwards Air Force Base).[2] The NACA engineers who came to the high desert north of Los Angeles, California, did so at a time when the first and second generations of experimental rocket planes were still central to the work done at the flight research outpost. In the course of their duties many of these engineers spent time calculating aerodynamic drag of unpowered aircraft because rocket planes were air-launched, expended their fuel in a brief flight, and landed without propulsive power. It was important to know what was happening in terms of lift and drag once a rocket plane ran out of fuel, and this included information about drag generated by the rocket plane's blunt, truncated afterbody, where a cluster of rocket nozzles was located. The pinnacle of the rocket plane program was the success of the X-15, the world's first hypersonic and space plane, on which many of these engineers worked. (By this time the agency had evolved into NASA.) Others eventually spent time on one or more of the family of lifting bodies: peculiar, wingless aircraft meant to explore the possibility of flying to a landing after returning to the atmosphere from space.

When designing aircraft, or even parts of aircraft such as external fuel tanks, aerodynamicists usually strive for as efficient a shape as possible. If speed and efficiency are the goal, blunt shapes are to be avoided.

Aerodynamicists usually want as streamlined a shape as possible, and this applies both to the fore and the afterbody of an object. In principle, the cleaner a shape the less drag it will generate—the shark might be regarded as the ideal body for passing through water, for example, with its sharp nose and a body that tapers to almost nothing.

But lifting bodies, rocket planes, and hypersonic vehicles share a fairly singular and seemingly contrary aerodynamic feature: a blunt afterbody. Although such a shape is undesirable for most aircraft, it is an unavoidable part of aircraft intended for atmospheric entry and an inevitable feature of rocket planes. The engineers at the Flight Research Center (FRC), as it had become known by the 1960s, wrestled with this characteristic not simply because it was counterintuitive, but because high aft-end drag had dramatic, sometimes positive effects on a vehicle's stability in flight. Indeed, that high drag was not merely an unavoidable reality but was often desirable on certain aircraft.[3] This experience became central to the exploration of truck aerodynamics pursued at the FRC in the 1960s and 70s.

Glancing at an image of the FRC's own design for an aerodynamic tractor-trailer, and then at an image of today's long-haul trucks, the heritage of the latter may not be obvious. One reason is that the FRC's truck was a cab-over, a style less and less often in use today for long-haul freight. Another is that, despite results gleaned through the center's research, few manufacturers adopted the suggested changes except to tractor cabs. Nevertheless, current long-haul trucks owe a great deal to empirical research conducted at the NASA center, and to the publications emanating from there over the years informing those interested in the pitfalls and gains that lay ahead.

---

[1] One only has to look at the agency's publication *Spinoff* or the list of patents held by the agency to grasp the extent to which this is true.

[2] The NACA was founded in 1915 in response to lagging performance in the American aeronautics field when compared to European advancements. Despite the Wright brothers having been the first to fly, Europeans soon eclipsed Americans in the field, a reality driven starkly home by the events of World War I. The agency was transformed into NASA in 1958, in direct response to the successful Soviet launches of Sputnik I and II.

[3] This bluntness is not restricted to the aft end of such vehicles. It appeared on the trailing edge of the ailerons and flaps of the X-2, for instance. The blunt trailing edges permit gradual sloping on the aft portion of these surfaces, which was beneficial at supersonic speeds since it improved control effectiveness.

2 Drag

A reflection on NASA's interest in aerodynamically efficient trucks turns out to be more than just an account of fairings and base drag and surface roughness. This history is also about technological choices, cultural values, and how Americans define themselves. And since technology is a reflection of human choices and values this comes as no real surprise.

Chapter Two
Eddies and Currents

While regularly riding his bicycle from his home in North Edwards to the NASA Flight Research Center (today's Dryden Flight Research Center) on Edwards Air Force Base, Edwin J. "Ed" Saltzman noticed the push and pull of tractor-trailers as they passed him. Saltzman's route took him along a section of Highway 58 in Southern California's High Desert before veering off toward the base. The Dryden Flight Research Center is one of several tenants of the U. S. Air Force at Edwards.[1] Highway 58 was, and is frequented by trucks coming from Arizona, Nevada, or Northern California that seek to avoid the greater Los Angeles area. As these tractor-trailers came upon Saltzman he first felt the bow wave of air pushing him away from the road and toward the sagebrush and tumbleweeds. But as the trucks swept past, their wakes had the opposite effect, tending to draw him toward the road, even causing rider and bicycle to lean into the lane. Anyone who's ridden a bicycle next to fast-moving traffic has felt some of this, although the full effect is available only to those bold enough to mingle with over-the-road tractor-trailers at highway speeds.

Saltzman came to the High Desert in 1951 to work as an engineer for the NACA, just four years after a human first successfully exceeded the speed of sound. He cut his aeronautical engineering teeth on the X-1 rocket planes that were still being used to explore the transonic and supersonic realms. In 1953, as the X-1 program wound down and the first Mach 2 flight took place, he began working on "Project

Edwin "Ed" J. Saltzman at his desk at the Flight Research Center. In his hands is the primary tool of the day for flight test engineers: a slide rule, this one likely a 20-inch model.
NASA   E58-3338B

Pilot Bill Dana stands in front of the North American Aviation X-15 following a successful landing on Rogers Dry Lake at Edwards Air Force Base.
NASA   E-1716

At left, *the Bell X-1, the first aircraft to exceed the speed of sound. Like most of the earliest X-planes, it was rocket-powered.*
NASA   E52-0670

4 Eddies and Currents

The Douglas D-558-2 being loaded into the belly of the mothership. Like the X-1, it was rocket-powered, and it was the first aircraft to exceed Mach 2.
NASA
E-1013

1226," later known as the X-15. The flight portion of the X-15 program lasted from 1959 to 1968, but it was preceded by years of engineering work to which Saltzman contributed.[1] As he had on the X-1 program, he worked as an aerodynamicist focusing on questions of lift-to-drag ratios (L/D). While the NACA's X-1 and D-558-2 aircraft featured blunt afterbodies because of their rocket motors, the X-15 dwarfed both in terms of the rocket nozzle area and the resulting drag the aft end generated.[2] Even before the X-15 program ended, Saltzman transferred to the XB-70 program, working on the Mach 3 experimental bomber that never saw production but which NASA used to explore high-speed atmospheric flight. Again, he served as an aerodynamicist.

In 1972, Saltzman began thinking about the rela-

---

[1] Visitors to the Smithsonian's National Air and Space Museum in Washington, D.C. can view one of the two remaining X-15s. Alongside a second-floor railing from which the X-15 is visible, a display case holds three technical papers, examples of research done with the X-15. Ed Saltzman is the author of one of those papers.

[2] The X-1s and D-558-2 all had four-chamber rocket motors located in the fuselage directly beneath the vertical stabilizer. The chambers were arranged in a diamond pattern. In turn, four rocket nozzles were similarly arranged, with the lip of each nozzle meeting the edge of a flat plate that formed the end of the airplane. In the case of the X-15 (in its final configuration), the single rocket motor had a diameter nearly as large as the aircraft's fuselage. Above and below it, both vertical stabilizers were unconventional in that they were wedge-shaped, lacking the traditional tapering of a flying surface's trailing edge. At subsonic speeds the base drag of the X-15 constituted more than 80 percent of the total drag at low-lift conditions.

tionship between his bicycle and trucks, thoughts stemming from his rides to and from work. Could a truck's bow wave and the trailing wake be reduced or mitigated?

Like any vehicle, a truck pushes air ahead of it as it moves, not unlike the bow wave of a boat with a flat prow: the greater the speed, the greater the bow wave. The amount of air being pushed is not insignificant; a modern (nicely faired), conventional tractor-trailer unit moving at 55 mph displaces as much as 18 tons of air for every mile it travels.[3] From an aerodynamicist's perspective, the bow wave is a localized high-pressure zone, for the air in that region is pushed forward by the truck's front surface.

Meanwhile, the opposite is developing at the back end of the trailer. As the truck rolls forward, air is pushed ahead, some of it moving around and over the cab, then unevenly down the side of the trailer. At the end of the trailer the displaced air is suddenly confronted with an abrupt 90° turn it cannot negotiate. Consequently, a low-pressure zone develops just behind the flat end (the base) of the trailer. Into this low-pressure zone eventually tumbles the chaotic airflow from the tractor and trailer, creating additional drag. The simple description belies a more complex activity, however. The combination of the low-pressure zone and the vehicle's forward motion actually cause the air swirling into the low-pressure zone to flow forward, in the same direction as the trailer. In effect, the trailer pulls this air (a portion of that 18 tons of displaced air) with it, and this takes energy. The high pressure at the front, turbid air alongside as well as on top and under the vehicle, and the low pressure at the back combine to generate considerable aerodynamic drag.[4] The shape of long-haul trailers is determined by function, just as an ocean-going shipping container's is: rectangular and cube-like to create the maximum amount of internal shipping volume. Yet even when the object's shape is, by necessity, one of the least aerodynamic, drag can be controlled or lessened with modifications to its external shape. This is what Saltzman contemplated, and he did so not only in the context

The North American Aviation's triple-sonic experimental bomber, the XB-70 Valkyrie, one of the projects on which Ed Saltzman worked during his Dryden career. Two XB-70s were built, but the model was never put into production. After one airframe was lost in a 1966 accident, NASA used the second for a series of supersonic experiments before retiring the aircraft to the U.S. Air Force Museum at Wright-Patterson Air Force Base in Dayton, Ohio.
NASA   ECN-1814

of his own knowledge of aerodynamics, but also in the crush of the first peacetime fuel crisis faced by the United States, in 1973.

Shapes

Modifying the shape of motor vehicles to improve their aerodynamics is not a new pursuit. Records of such attempts to improve aerodynamic efficiency date to the early twentieth century, when land vehicles (other than trains) regularly began exceeding the speed of a horse. A German designer, for example, drew up plans in 1914 for a highly faired car, rounded at the front and tapering to a point at the tail. It even had round windows in a nod to portholes. In 1920 Cornelius Meyers received a U.S. patent for an invention the shape of which he expected to "furnish the vehicle with a novel air deflector which will prevent the for-

---

[3] *Automotive Engineering* August 1975, vol. 83, no. 8, pp. 40-43 (n.a., but the article is drawn from *Society of Automotive Engineers* papers by P. Lissaman (750702), and William T. Mason, Jr. (750707).

[4] Vortices that develop as the air separates from the trailer's surface are the problem: the turbulent swirls of air, sometimes referred to as a jet pump, actually draw air away from the base, dropping its pressure, and this lower pressure creates additional drag, called "base drag." In effect, the truck has displaced some 18 tons of air once, and now must pull some it along, expending even more energy.

mation of eddies of air at the rear end of the vehicle." The device had the added benefit of being collapsible, since in its functional state it extended aft of the vehicle by a considerable margin. In 1936 Edward A. Stalker received a U.S. patent for a modified automobile. He envisioned a sophisticated gear-driven suction pump that drew air in through two slots at the back of the car (in effect, an early effort at boundary-layer control) to minimize, if not prevent, flow separation.[5] By the early 1950s, there had been theoretical as well as some limited experimental work conducted on the subject of ground vehicle aerodynamics; some of the latter even entailed use of models in wind tunnels. And in 1956, R. D. Potter registered an "inflatable streamlining apparatus for drag reduction," one of several patents from the period for aerodynamic devices targeting trucks.[6] Indeed, by the 1960s several individuals had proposed streamlining long-haul trucks, although little of the work underlying their patent applications was more than theoretical.

Patent records alone show that through this period more than a few ideas emerged for improving the aerodynamics of vehicle efficiency, particularly that of trucks. When Saltzman and his colleagues began their research it was not as though they were the first to address the issue. There were differences this time, however. For one, these were practicing aeronautical engineers with experience on the very subject, not theoreticians or wind-tunnel researchers or people

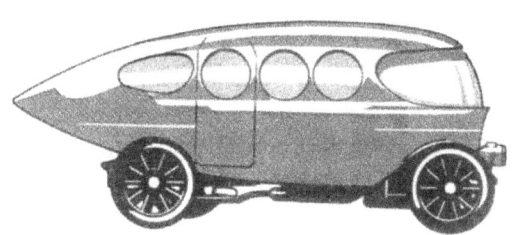

This illustration of an aerodynamic car dates from 1914.

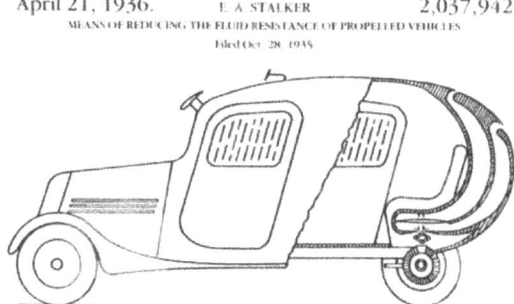

E.A. Stalker's 1936 patent drawing for a car with boundary layer control.

---

[5] Edward A. Stalker was a professor of aeronautical engineering at the University of Michigan, and served as that department's first chair. Among his students was Clarence "Kelly" Johnson, who went on to make a name for himself with Lockheed Aircraft. See: Barnes McCormick, Conrad Newberry, and Eric Jumper, eds., *Aerospace Engineering Education During the First Century of Flight* (Reston, VA, American Institute of Aeronautics and Astronautics, 2004), 48-50, and http://www.wvi.com/~sr71webmaster/kelly1.htm (accessed 28 December 2009).

[6] See for example Cornelius T. Meyers, "Air-Deflecting Device," U.S. Patent No. 1,352,679; E. A. Stalker, "Means of Reducing the Fluid Resistance of Propelled Vehicles," U.S. Patent No. 2,037,942, April 1936; E.A. Dempsey, "Vehicle Body and Attachment Therefor," U.S. Patent No. 2,514,695 of July 1950; R. D. Potter, "Inflatable Streamlining Apparatus for Vehicle Bodies," U.S. Patent No. 2,737,411, 6 March 1956; Alexandre Favre, "Aircraft Wing Flap with Leading Edge Roller," U.S. Patent No. 2,569,983, 2 October 1951; A. F. Stamm, "Tractor-Trailer Airstream Control Kit," U.S. Patent No. 2,863,695, 9 December 1958; Walter Selden Saunders, "Drag Reducer for Land Vehicles," U.S. Patent No. 3,397,120, 10 October 1972; Neal A. Cook and Gerard Friedenfeld, "Vehicle Space Closing Means," U.S. Patent No. 3,834,752; Ronald A. Servais, "Streamlining Apparatus for Articulated Road Vehicle," U.S. Patent No. 3,945,677, 23 March 1976 ; Edgar L. Keedy, "Vehicle Drag Reducer," U.S. Patent No. 4,142,755, 6 March 1979; A. Wiley Sherwood, "Wind Tunnel Test of Trail-mobile Trailers," Wind Tunnel Report No. 85, University of Maryland (June 1953); Donald S. Gross, *Wind Tunnel Tests of Trailmobile Trailers*, 3rd series /prepared by Donald S. Gross (College Park, MD: University of Maryland, College of Engineering, Glenn L. Martin Institute of Technology, Wind Tunnel Operations Dept., [1955]); H. Schlichting, "Aerodynamic Problems of Motor Cars," AGARD Report 307 (October 1960); Harold Flynn and Peter Kyropoulos, "Truck Aerodynamics," Society of Automotive Engineers Transactions (1962), vol. 10, pp. 297-308; Sighard F. Hoerner, "Fluid-Dynamic Drag," Midland Park, N.J: by the author, 1965; J. W. Anderson, J.C. Firey, P. W. Ford and W. C. Kieling, "Truck Drag Components by Road Test Measurement," Society of Automotive Engineers Transactions (1965), vol. 73, pp. 148-159, 186; G. W. Carr, "The Aerodynamics of Basic Shapes for Road Vehicles, part 1—Simple Rectangular Bodies, Report no. 1968/2," Motor Industry Research Association (November 1967); Gary L. Smith, "Commercial Vehicle Performance and Fuel Economy," Society of Automotive Engineers SP-355 (January 1970). For a larger sampling of early U.S. patents for aerodynamic vehicles, see Appendix A.

with random ideas.[7] For another, recent fuel shortages and price increases made the results relevant, whereas earlier works had no similar undercurrent.

The trucking industry, at which nearly all the efforts of the 1950s were aimed, seemed indifferent to the searches for greater fuel efficiency since the industry "did not consider the fuel savings by these modifications significant. The suggested devices were not considered practical," wrote Vincent Muirhead and Ed Saltzman in an article for the Journal of Energy.[8] Looking back at the patents of the period, skepticism is a reasonable sentiment, for while the proposed shapes appeared potentially beneficial, the ideas on which they were based had little grounding in aerodynamics and almost no empirical evidence to back the claims inventors made for them. Moreover, so long as fuel prices remained low, few seemed interested in what these modifications might offer. Instead, the rising aerodynamic drag associated with increasing truck speeds "was merely overcome by more powerful engines," a time-tested solution. As horsepower climbed so, too, did fuel consumption, but trucking firms and owner-operators did not seem to mind; fuel was cheap.[9]

Aerodynamically speaking, the shape of a truck will not matter significantly until it can reach a certain speed. For much of the early twentieth century, truck manufacturers sought enough power to match the growing loads of the trucks; the trucks themselves weren't capable of speeds at which aerodynamics became a factor. It was only when trucks were powerful enough to pull their full load and exceed a certain speed that their shape became an issue. A parallel exists in aviation: retractable landing gear does not matter much on airplanes flying slower than 250 mph. Adding such gear marginally reduces drag on an airplane below that speed, but it also adds weight and complexity; as a result, there is no net gain. Only when crossing this approximate speed threshold does the net aerodynamic gain of retractable landing gear outweigh the cost.

It would take an outside force to shift thinking within the trucking industry, and that force came in 1973 with the first non-war-related fuel crunch of the nation's history.[10] Enter Saltzman and the small group of NASA engineers at the FRC in the fall of that year, conscious of the fuel crisis gripping the country and, most important, sensitive to the impact of aerodynamics on moving vehicles, particularly bluntly shaped ones.

The Foundation

In the early 1960s the Flight Research Center began testing the first full-scale lifting body, the M2-F1. Conceived to offer an alternative to ballistic capsule atmospheric entry from space, a lifting body is designed to glide to a landing once inside the atmosphere

The M2-F1 on tow behind a C-47 over Rogers Dry Lake. The first lifting body, it was built of wood with a steel internal truss that held an ejection seat and landing gear. Clearly visible from this angle is the vehicle's blunt base, which had an L/D ratio of 2.8:1.
NASA   E-10962

---

[7] This in no way impugns the work of value by academics, on whom practicing engineers rely in a multitude of ways. But there are distinctions in opportunities that benefit both parties differently.

[8] Vincent U. Muirhead and Edwin J. Saltzman, "Reduction of Aerodynamic Drag and Fuel Consumption for Tractor-Trailers Vehicles," *Journal of Energy* vol. 3, no. 5 (September-October 1979): 279.

[9] Louis L. Steers, Lawrence C. Montoya, *Study of Aerodynamic Drag Reduction on a Full-Scale Tractor-Trailer* (National Aeronautics and Space Administration, Dryden Flight Research Center, Edwards, CA) in conjunction with the US Department of Transportation, Washington, D.C. DOT-TSC-OST-76-13 (National Technical Systems Service, Springfield, VA, 1976).

[10] A worldwide energy crisis began in late 1973 when, following the end of the Arab-Israeli War of that year, members of OPEC (the Organization of Petroleum Exporting Countries) imposed an oil embargo on much of the world, leading to a quadrupling of oil prices in very short order.

The back end of the Northrop M2-F3, highlighting the blunt aft end and the rocket nozzles.
NASA   E-21533

The Northrop HL-10, seen from the rear, before the rocket motor had been installed (beneath the center vertical stabilizer). This angle illustrates the blunt base so typical of lifting bodies, but does not reveal the amount of increased drag produced when control surfaces are deployed on descent (re-entry). In that event, inboard and outboard surfaces of all three vertical surfaces deploy, as do the top and bottom "flaps" on the body, dramatically increasing drag.
NASA   ECN-1463

rather than float back to Earth beneath a parachute.[11] Over the span of that decade, the FRC flew a series of lifting bodies to explore the concept. Lifting bodies built in the follow-on series were considerably heavier than the M2-F1, and carried a rocket motor to provide added speed and altitude during flight, the better to mimic a return from space. Their shapes varied slightly with each successive model but one feature remained constant: a blunt aft end. This was one of the few characteristics common to the lifting bodies and earlier generations of rocket planes. Consequently, the aerodynamics of vehicles' blunt aft ends was something Saltzman and other FRC engineers knew quite a bit about.

As exo-atmospheric vehicles, both the lifting bodies and the X-15s were designed to fly at least a portion of their descent at hypersonic speeds, and in both cases a defining characteristic of these aircraft was their blunt aft end.[12] The reason for the abrupt aft end (in addition to the area of the rocket nozzle) lies in the increased stability provided by the high drag of such a shape. The X-15's vertical stabilizers, for instance, were wedge-shaped so that their blunt bases generated lateral stability at key speeds. In the case of the lifting bodies' blunt aft end brought stability to aircraft that often had marginal flying qualities. The drag that came with this would, in theory, slow the lifting body, enabling it to negotiate atmospheric entry from orbit without special materials and without burning up.

---

[11] M referred to "manned" and F referred to "flight" version, as opposed to a wind-tunnel model, hence the M2-F1 was the second man-rated design, but the first to be built for actual flight. For more on the lifting body program see R. Dale Reed with Darlene Lister, *Wingless Flight: The Lifting Body Story* (Lexington: the University of Kentucky Press, 2002).

[12] The only lifting body to have exited and entered Earth's atmosphere was the X-23 PRIME, built as a maneuvering reentry test vehicle. The first vehicle was launched on the top of an Atlas rocket from Vandenberg Air Force Base on 21 December 1966. It was lost during reentry in the Pacific Ocean. Of the three X-23s, only one was recovered in flight while suspended from its parachute stringers, as planned. The lifting bodies referred to in this manuscript were designed to explore and validate this possibility. The highest and fastest lifting body, the HL-10, managed Mach 1.86 and 90,020 feet altitude on two separate flights. The X-15, on the other hand, flew 199 times, 13 of these flights to space and back.

This photo of the North American Aviation X-15 shows the aircraft's blunt aft end, a result of the two vertical stabilizers, the fuselage's circumference, and the rocket engine area. This is the aircraft's initial configuration, before the XLR-99 engine had been fitted; two XLR-11s were being used instead, but the same base area remained.
NASA   E-5256

Chapter Three
The Shoebox

Curious as to what might be done to improve the aerodynamics of a blunt vehicle on land, Saltzman initially coaxed fellow engineer Victor W. "Vic" Horton into using the latter's pickup truck for the first experiments. Horton, like Saltzman, had worked on the M2-F1, and so had considerable experience with blunt afterbodies of aircraft, and he also served as a flight test engineer for NASA. The two men assembled a rudimentary set of instruments to use in establishing a base drag for Horton's pickup so that Saltzman could approach the center director with a formal request for financial support.[1] After collecting data during several runs with the truck, Saltzman and Horton went to Milton O. "Milt" Thompson, then chief of research projects at the center, and director of research Joseph Weil with a modest proposal, hoping to add Thompson's and Weil's endorsements to a presentation for the center director. After reading the proposal (to which he gave his full support), Thompson appended a note, dryly commenting: "the results of this [research] should be so obviously productive that it probably won't get approved."[2] The quip notwithstanding, Lee R. Scherer, then center director, agreed to fund the project, and Saltzman soon arranged to use the center's old mail delivery van as the first formal test bed.[3]

The test vehicle chosen was a Ford passenger van that had been retired from delivery duties at the center.[4] The small group of engineers gathered by Saltzman for the project began by first establishing the van's baseline drag: its tractive and aerodynamic drag. Identifying the tractive drag meant accounting for resistance in the driveline, u-joints, wheels, and tires. Preexisting data for the friction drag of tires, generated primarily by tire manufacturers, was factored into the equation. But tire friction changes with velocity, and this, along with details such as tire pressure and vehicle

The NASA delivery van that the center turned over to Saltzman and his team for use in the first set of formal tests in aerodynamic research on land vehicles.
NASA   *E73-26449*

weight—even the rolling inertia of the wheels—had to be accounted for.

Despite the existing data, establishing the baseline drag of the van ultimately required that it actually be put in motion, so engineers took the van to the Edwards South Base runway, the base's first paved runway that was, by then, used almost exclusively by the Edwards flying club. The first step in determining the tractive drag was to attach a "fish scale," a simple, large, spring-loaded scale, to the van's front bumper. With the van parked on a flat surface, its brakes off and the transmission in neutral, one of the team slowly pulled on the scale to move the vehicle forward. Whatever force in pounds registered on the scale represented the tractive drag of the van. The rolling friction of the van, the force it took to keep the van in motion, was 45 and 50 lbs. This exercise was conducted before and after each day's run, just as were weighing the

---

[1] V. W. Horton, R. C. Eldgredge, and R. E. Klein, *Flight-Determined Low-Speed Lift and Drag Characteristics of the Lightweight M2-F1* (Edwards, CA: NASA TN-D3021, 1965).

[2] Memo from Milt Thompson to Ed Saltzman, in the personal collection of Ed Saltzman.

[3] Memorandum to Director regarding "Request for approval of research and development project—study the increase of efficiency of ground vehicles," from E. J. Saltzman and R. R. Meyer, 22 November 1972.

[4] Their first experiment employed Vic Horton's pickup truck with a camper shell on the bed.

van and checking the pressure in all four tires, all to ensure that that day's data had a consistent baseline. Any variation between one day's numbers and another's—a drop in tire pressure during the day's runs, for instance—voided the data for the entire day. Despite the number of variables, the figures were remarkably consistent and no day's work was thrown out over a sudden change in tractive drag.[5]

Establishing aerodynamic drag meant using the "coast-down method," described by Sighard F. Hoerner in his book *Aerodynamic Drag* but in use by others before even he wrote about it.[6] The FRC engineers had already developed considerable experience with this method working on the X-1 and X-15 programs (indeed, in work with any rocket airplane), since both aircraft were powered by a finite amount of propellant and once that was consumed in flight, the aircraft became a large, heavy glider. Those airplanes were thoroughly instrumented with accelerometers that measured deceleration, which was translated into drag once the propellant burned out. (The researchers did briefly apply accelerometers to the van as a check on the "speedometer-stopwatch" method.)[7]

To determine the total vehicle drag, a team member started the van and accelerated down the runway to a predetermined speed (usually 65 mph), took the van out of gear, and coasted to a low speed, typically 25 mph. Using a series of stopwatches, engineers measured how long it took for the van to reach successively lower speeds, making sure to conduct the test going in both directions on the runway to negate wind effect and any material incline in the runway. The team repeated this several times to develop an average. But how to separate the aerodynamic drag from the total drag that emerged using the coast-down method? Knowing the mechanical drag of the van, and able to factor in both rotational and inertial forces at play, they were left with only aerodynamic drag. But they repeated the coast-down test multiple times to ensure consistent data.

Nevertheless, anxious to be sure their data were reliable and repeatable, the engineers attached a small square plate to a structure that was itself attached to the van's roof, with the flat side facing forward. The dimensions and drag data for the plate and attaching structure came from Hoerner, so the engineers knew the amount of aerodynamic drag they were adding to the van. They then ran a test and measured the amount of new drag they had introduced, which corresponded to their own predictions (and the amount indicated by Hoerner), after which they were sure of their initial calculations.[8] This then served as the baseline drag of the unadulterated van throughout the experiments.

At this point the van was taken to the center's shop, where mechanics began work on it. Following engineering drawings, the mechanics built an aluminum frame around the vehicle and attached aluminum sheets to the framework. Where the windshield, driver, and passenger side windows were, they affixed sheets of Plexiglas. And they attached louvers to the front of the van that could be operated from within, which allowed or prevented airflow directly to the radiator. The louvers were opened when the van was in normal operation but always closed during the coast-down intervals. In the new van's first configuration, all the edges formed 90° angles and it lacked mirrors and lights of any kind. The van now resembled an aluminum shoebox on wheels, and was soon given this moniker by project members.

---

[5] These same tests, including the "fish scale" test, were conducted on the cab-over-engine trucks as well once the group began testing vehicles of that size.

[6] Sighard F. Hoerner, *Aerodynamic Drag: Practical Data on Aerodynamic Drag, Evaluated and Presented by Sighard F. Hoerner* (Midland Park, NJ, 1951). Hoerner methodically established drag coefficients for various shapes and sizes of objects, from flat plates to round objects. It is possible to extrapolate drag data simply by using his tables, but it was also necessary for team members to assemble additional data on rolling drag in order to develop a complete picture about the box van.

[7] For more detail on developing the tests, see Edwin J. Saltzman and Robert R. Meyer, Jr., *Drag Reduction Obtained by Rounding Vertical Corners on a Box-Shaped Ground Vehicle* (Edwards, CA: NASA-TM-X-56023, 1974), Edwin J. Saltzman, Robert R. Meyer, and David F. Lux, *Drag Reduction Obtained by Modifying a Box-Shaped Ground Vehicle* (Edwards, CA: NASA, TM X-56027, 1974), and also Lawrence C. Montoya and Louis L. Steers, *Aerodynamic Drag Reduction Tests on a Full-Scale Tractor-Trailer Combination with Several Add-On Devices* (Edwards, CA: NASA TM X-56028, 1974).

[8] Saltzman to Gelzer, notes on manuscript draft. See also Saltzman and Meyer, *Drag Reduction Obtained by Rounding Vertical Corners on a Box-Shaped Ground Vehicle*. The van had a manual, three-speed transmission that was placed in neutral for the coast-down tests. At that point the tractive drag consisted of gear resistance down the driveline to the differential and axle, tires, bearings, combined with rotational inertia.

Following the calculation of tractive drag, center researchers attached this fixture, the aerodynamic drag of which Sighard Hoerner had already determined, to the van, in an effort to validate their predictions for the van's aerodynamic drag. It was removed after this lone test, having proved the predictions correct.
NASA   E73-26454

The Ford mail delivery van with its initial substructure attached. To this the fabrication shop attached sheet aluminum.
NASA   E73-26478

The engineers set about running tests of the square-cornered Shoebox using the coast-down method. To the set of stopwatches used to determine the length of time needed for the van to decelerate, they added an accelerometer and a recording oscillograph, as a functional as well as a backup data device. The first tests of the modified van complete and data in hand, engineers sent the Shoebox back to the shop for further modification. This time technicians rounded the vertical edges as well as all four vertical corners. Again, they took the van out for coast-down tests on the South Base runway. The contrast between the two configurations was eye-opening. At 55 mph the Shoebox with rounded corners had the same aerodynamic drag as did the square-cornered Shoebox at 44 mph. A second test showed that the rounded-corner Shoebox had the same aerodynamic drag at 70 mph as the square-cornered Shoebox had at 55 mph.[9] The team's coast-down measurements became more refined in the process; now, as many as six stopwatches were operated by the passenger in the van to measure the time intervals needed for the van to decelerate in 5-mph increments.[10] Each day, the team towed the van to the base scales before each run, getting official weights for each trip, then passed through the scales again on the way back to the center after the day's runs.[11] (By the time their research project was complete, Saltzman's team had become the scales' best customer.) Tests led to new modifications, followed by more tests, followed by more modifications. Then, once the team had finished modifying the van's exterior shape, they moved to the underbody, sealing it entirely, including the wheel wells.[12] Rounding all four vertical edges and corners yielded a reduction in aerodynamic drag of 54 percent, while sealing the bottom of the vehicle reduced drag another 15 percent, for a cumulative 61 percent reduction in aerodynamic drag over that of the Shoebox's original configuration. (As a measure

---

[9] Saltzman and Meyer, *Drag Reduction Obtained by Rounding Vertical Corners on a Box-Shaped Ground Vehicle*, 5.

[10] Ibid.

[11] Lacking mirrors or lights of any kind, the Shoebox could not be driven on even the base roads.

[12] Saltzman, Meyer and Lux, *Drag Reduction Obtained by Modifying a Box-Shaped Ground Vehicle*, passim.

14   The Shoebox

The image at left shows the underbody of the Shoebox in its stock configuration. The image at right shows the same underside after it had been sealed. This modification alone, including sealing the wheel wells, reduced aerodynamic drag by 15 percent.
NASA   *E74-27716, E74-27626*

of how small changes can make big differences, not rounding the rear corners cost 5 percent in drag reduction.) The engineers estimated the potential gain in fuel economy for the aerodynamic improvements to be between 15 and 25 percent at highway speeds, and for very little loss of internal volume.[13]

Perhaps just as significant as the raw data was this: in its original configuration, the Shoebox "had a sig-

As the tufts of yarn make clear, applying a radius to all corners of the Shoebox, front and back, smoothes airflow.
NASA   *E74-27068*

---

[13] Ibid., 7; also Randall L. Petersen, *Drag Reduction Obtained by the Addition of a Boattail to a Box Shaped Vehicle* (Edwards, CA: NASA CR-163113, August 1981), 1.

nificantly higher drag coefficient than generally similar small scale models" used in wind tunnels. The team attributed this to the underbody protuberances that were not reproduced on the models. The importance of this remark, almost lost in the conclusion's mound of drag reduction figures that were the focus of the report, would become evident barely a year later, when the trucking industry began to pay close attention to the center's research on truck drag reduction.[14]

There are two reasons why the correlation between aerodynamic drag reduction and fuel gains for a rolling vehicle does not constitute a one-to-one ratio. First, total drag is a function of multiple factors, only one of which—aerodynamic drag—can be controlled by reshaping the vehicle. Second, because of friction and pressure, the faster an object moves through a fluid the more drag it generates. Though velocity increases linearly, drag, which is a function of velocity (V), goes up by the square of the velocity increase (V^2). Going from 55 mph to 65 mph is an 18 percent increase in speed, but the aerodynamic drag jumps by 40 percent for that same increase. But that's not the worst of it: the power to overcome that jump in drag increases by the cube of velocity, or V^3. (In the case of tractor-trailers, roughly half the truck's horsepower is needed simply to overcome aerodynamic drag when traveling at 55 mph.)[15]

The engineers were not dismayed by any of this, however, in part because they knew full well the disparity between reduction in aerodynamic drag and reduction in fuel consumption. More to the point, the Shoebox was only the starting point, a preliminary test bed. After all, Saltzman had not been mingling with delivery vans on his way to work.

---

[14] Saltzman, Meyer and Lux, *Drag Reduction Obtained by Modifying a Box-Shaped Ground Vehicle*, 10.

[15] John Allen, a British aerodynamicist, recounted changes the Volkswagen Company made to its initial Kombi minibus in the early 1960s. "The original design had sharp front edges, and tufts along the sides showed the flow completely broken away and turbulent. Even quite modest rounding of all the front edges and corners streamlined the airflow and reduced drag by 40%. The fuel saving on all Volkswagen buses in service at this time corresponds to 130,000 tons per annum." John E. Allen, *Aerodynamics: The Science of Air in Motion* (London: Hutchinson & Co., Ltd, 1963), 2nd ed., Granada Publishing, 1982, 90. Researchers at the FRC knew of this work but chose to proceed for several reasons. First, they wanted to vary other aerodynamic factors, such as was accomplished by sealing the underside of the vehicle; second, they believed that NASA publications would be more far-reaching; and third, they wanted to experiment on real vehicles rather than on wind-tunnel models.

Chapter Four
Word Spreads

Even before the group published its first report on the Shoebox, word about the research began to ripple beyond the center. At the beginning of 1974, while the team was still writing its first report on the aerodynamic van project, the U.S. Department of Transportation approached the Flight Research Center with the idea of testing the effectiveness of several add-on, aftermarket products that claimed to improve fuel efficiency for long-haul trucks of the cab-over-engine design, the type most common in that period.[1] The DoT offered funding to NASA to test five devices.[2]

FRC managers accepted the offer, and soon after engineers prepared a series of tests applying the same methods used on the test van, with each device to be carried on a truck driving up and down the unused runway at South Base, measuring the drag reductions—if any—for each modification.

Two of the five products had already been on the market for several years. The first, identified by the letter A in the study, was an Airshield, made by the Rudkin-Wylie Corporation of Connecticut. By October 1974, the Airshield had been available for six years and the company claimed to have sold some 24,000 units. Although the device had a favorable reputation it had not been evaluated by an impartial agency, such as NASA.

The Airshield consisted of a roof-mounted air dam on the cab and a plate called the Vortex Stabilizer, which ran perpendicular to the air dam but was positioned behind it, filling some of the gap between cab and trailer. In a tractor-trailer combination the Vortex Stabilizer was attached to the trailer's front face, and on the single-chassis truck the two were attached so as to form a "T" with the cab roof.

The other product that had also been on the market for some time came from FitzGerald Nose Cone, a firm in California, identified with the letter C in the

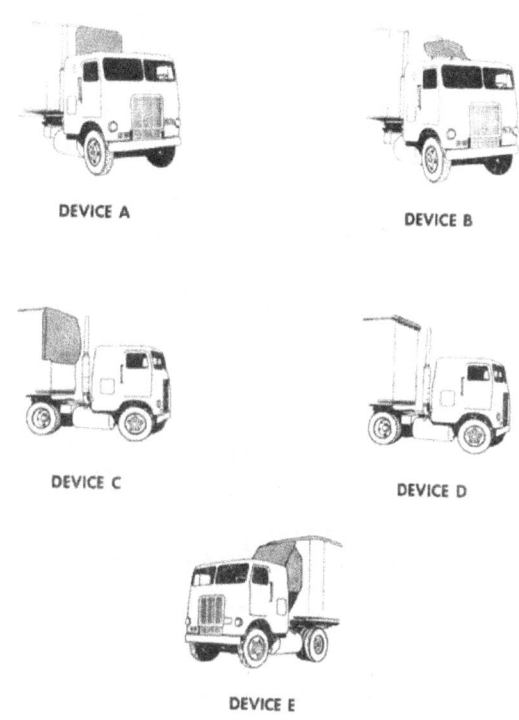

The five aftermarket devices the Department of Transportation contracted with the Flight Research Center to test on a tractor-trailer, identified by the darkened area in each illustration.
NASA/U.S. DoT

study. FitzGerald Nose Cone had its origins in the 1960s when its founder, Joseph FitzGerald, worked for Carrier TransiCold, a supplier of refrigeration units for truck trailers. In 1965 FitzGerald suggested moving the evaporator unit from within the trailer

---

[1] The tractor used in the tests was a 1974 Freightliner cab-over-engine with a sleeper compartment, powered by a Cummins 320-hp diesel engine. The tractor pulled a two-year-old 45-foot trailer made by Strick.

[2] February 20, 1974, Request for Project Approval, Statement of Work for Joint DOT-NASA Truck Aerodynamic Study, 1/7/74: "The proposed study will primarily investigate the effects of existing add-on devices on truck aerodynamic drag and fuel consumption." The engineers submitting the proposal were L.(awrence) C. Montoya, Louis Steers, Bruce Powers, and Larry Reardon. From the personal collection of Edwin J. Saltzman.

to the nose of the trailer to provide more room for cargo. The company did so, and drivers noticed that the newly configured units were easier to drive and burned less fuel than the previously configured ones. FitzGerald ascribed this, correctly, to the relocated refrigeration unit, which, in its new position on the nose of the trailer, both improved the drag characteristics of the trailer and narrowed the gap between cab and trailer, serving to reduce crosswind influences on the combination.[3] By 1973 FitzGerald had started his own company to market blunt, slightly rounded pods that attached to the box of the trailer just above the cab roof (in case the truck was not a refrigerated unit). The firm offered the pods to short- and long-haul truckers and freight companies alike, promising improvements in fuel mileage. The company continues in operation as of this book's printing.[4]

The remaining devices FRC engineers tested, and the corresponding letters used to represent them in the study, consisted of a small, plow-like roof-mounted air dam (B), an air dam that actually had side panels to help seal the area between the cab and trailer (E), and a louver attached to the trailer's top front edge (D). This louver resembled a Handley Page automatic wing leading edge high-lift slat, which was meant to capture and redirect airflow over a wing. Unlike the HP slat that deployed forward of the wing or retracted flush with the wing depending on the aircraft's angle of attack (hence the device's automatic nature), article D was non-moving.[5]

Applying the same techniques they had on the Shoebox, Saltzman, Lawrence Montoya, and Louis Steers conducted a series of coast-down tests to establish a baseline drag for the standard cab-over-engine tractor-trailer, which they found had a coefficient of drag (Cd.) of roughly 1.06.[6] They then repeated the

NASA engineer Lawrence "Larry" C. Montoya
NASA　　　　　　　　　　　　　　*EC76-5688*

NASA engineer Louis L. Steers.
NASA　　　　　　　　　　　　　　*EC79-11038*

---

[3] NoseCone corporate history, http://www.nosecone.com/about.htm (accessed 3 June 2009).

[4] A recent advertisement for the company shows a model of a small trailer in a wind tunnel of sorts. In the ad, as wind increases on the front of the trailer, the trailer slides inexorably backward only to be pushed forward by a hand that suddenly appears. Once released, the trailer gradually slides backward again in the wind. Once the same trailer is outfitted with a Nose Cone fairing and placed in the wind tunnel under the same conditions, it sits firmly and does not move. http://www.nosecone.com/aepull.htm (accessed 31 December 2009).

[5] The three other products tested were: Airflo, Airvane, and Aerovane, made by the Airflow Company; Systems, Science, and Software; and Aero Van, respectively.

[6] Lawrence C. Montoya and Louis L. Steers, *Study of Aerodynamic Drag Reduction on a Full Scale Tractor-Trailer Combination with Several Add-On Devices* (Edwards, CA: NASA TM-X-56028, 1974). A version of this paper was presented in 1975 at a Society for Automotive Engineering meeting and published in the proceedings. See *Society of Automotive Engineers* 750703 (Warrendale, PA: Society of Automotive Engineers, 1975). NASA TM-X 56028 was, with the addition of further information derived from subsequent testing, also published by the Department of Transportation, in 1976. See L. L. Steers and L. C. Montoya, *Study of Aerodynamic Drag Reduction on a Full Scale Tractor-Trailer* Report No. DOT-TSC-OST-76-13 (Washington, D. C.: U.S. Department of Transportation, April 1976).

The NASA tractor-trailer carrying various aftermarket devices in a test for the Department of Transportation. Diatomaceous earth is blown through a tube up the cab's front and released between the three running lights at the cab's leading edge, revealing airflow for each device. The cab has also been tufted to show airflow around it while in motion. The bottom image shows the cab with a valance, applied by the FRC and unavailable as an aftermarket item.
NASA   EC74-4210

same tests with each aftermarket device. In order to assure as much of a constant rolling drag as possible with the test vehicle, the team filled the tires with nitrogen instead of air, minimizing variation in the tires' inflation because of heat. As with the Shoebox tests, once determined, they kept mechanical drag a constant in order that modifications to the unit (applying the add-on devices) produced changes only in aerodynamic drag.

One of the new tests the group performed was to vary the gap between the trailer and tractor from between 62 and 40 inches in an effort to see what effect this

NASA   EC74-4211

The third image in the sequence of the NASA tractor-trailer carrying various aftermarket devices in a test for the Department of Transportation.
NASA   EC74-4212

had on efficiency and drag. Data from the unmodified trailer runs showed that the larger the gap, the greater the drag. Simply moving the trailer forward to narrow the gap to just 40 inches netted a 7 percent drop in aerodynamic drag when traveling at 55 mph.[7]

Filling the gap between tractor and trailer might at first have seemed an easy solution to drag, but doing so was, and is, complicated by national transportation laws. Manufacturers could bunch the tractor and trailer more closely together, but this came with a penalty even while making it possible to improve efficiency: the further apart the two were, the greater the payload could be because the total weight could be spread over more road area. West Coast trucks typically had the largest gaps because they sought to carry the maximum load possible over the longest and lest uninterrupted hauls, yielding greater efficiency. The term "bridge formula" is used to describe this arrangement.[8]

To better understand the airflow around the tractor and trailer while both are in motion and sporting the devices, the group also conducted flow-visualization tests in which diatomaceous earth in powder form was released at the top front edge of the cab. As the powder billowed up and around the fairing and trailer, it illustrated where the air flowed while the truck drove at 55 mph. In the end, results showed that device A (Rudkin-Wiley) yielded between 2-4 percent drag reduction, device D (System, Science, and Software) only a 2-3 percent reduction. Device E, Aero Van Incorporated's Aerovane, not only deflected some of the air but also served to close the gap between tractor and trailer; it yielded an aerodynamic drag reduction of 19 percent. When a crosswind was factored into the tests the percentages dropped, but an improvement was still

---

[7] Lawrence C. Montoya and Louis L. Steers, "Aerodynamic Drag Reduction Tests on a Full-Scale Tractor Trailer Combination with Several Add-On Devices," (n.d.) (NASA Flight Research Center and the DOT Transportation Systems Center), 4-8. In contrast to the baseline tests of the unmodified trailer, devices A and B yielded the greatest improvement in drag with the trailer at the 62-inch gap, not the smaller, 40-inch gap, that best suited the unmodified trailer. The trailer gap made almost no difference for devices C and D. Device E effectively eliminated the gap altogether.

[8] "The bridge formula calculates the maximum allowable load (the total gross weight in pounds) that can legally be imposed on the bridge by any group of two or more consecutive axles on a vehicle or combination of vehicles. The bridge formula reflects the fact that loads concentrated over a short distance are generally more damaging to bridges than loads spread over a longer distance. It provides for additional gross weight as the wheelbase lengthens and the number of axles increases." http://training.ce.washington.edu/WSDOT/Modules/04_design_parameters/bridge_formula.htm (accessed 31 December, 2009). See also Paul Schenck, "New Focus on Air Drag," in *Trailer/Body Builders* November 1975, 37.

evident when compared to rates with a stock trailer.[9]

The next step was to test the fuel consumption of the tractor-trailer in service conditions. The team arranged for use of another tractor of the same type, also with DoT funding, and a trailer nearly identical to the one the FRC truck pulled. Ascertaining that both vehicles shared virtually the same aerodynamic and tractive drag (the coast-down method and a 300-mile road test found only a 1 percent difference between the two in fuel consumption), the team sent both vehicles out on the road for 312-mile, closed-loop road tests, with only one sporting the aftermarket drag-reducing devices. The trucks used the same gears for inclines encountered on the way, a coordination enabled by radio communication between the drivers, who also matched engine starts and stops as well as the opening and closing of windows and even air vents. There was very little traffic on the selected route through the Mojave Desert, all but eliminating other vehicles as a variable. The NASA engineers compared only the two devices (A and E) showing the greatest drag reduction on these road trips, and to obtain the best results they had the trucks driven over the same course three times with each device. In each case fuel was measured before loading, and then measured after the runs (weighing each tankload after allowing the trucks to cool down for a night, to eliminate density change as a factor in determining fuel remaining). The results matched the group's predictions based on the initial tests done on the South Base runway: the greatest drag reduction they identified resulted in an approximately 10 percent reduction in fuel consumption afforded by device A, with a slightly lesser dividend from device E of 9.3 percent.[10] The tests were complete by the summer of 1974 and they began publishing the results in October of that year.

Saltzman, Montoya, and Steers attended a transportation industry conference in October 1974, and in addition to a paper presented by Montoya they took the opportunity to distribute the report of their tractor-trailer tests. Later in the day of the presentation, at the motel where they were staying, the three discussed the industry representatives' reception of their tests and results, which each had noted as uniformly skeptical.

DEVICE A

The two best aftermarket devices to be taken out on road tests were A and E.
NASA/U.S. DoT

DEVICE E

---

[9] Lawrence C. Montoya and Louis L. Steers, *Study of Aerodynamic Drag Reduction on a Full Scale Tractor-Trailer Combination with Several Add-On Devices*, delivered to an SAE meeting in 1974, and Edwin J. Saltzman, "A Summary of NASA Dryden's Aerodynamic Truck Research" to be presented to the 1982 SAE Truck and Bus Meeting and Exposition, Indianapolis, IN, 8-11 November 1982.

[10] Louis L. Steers, Lawrence C. Montoya, and Edwin J. Saltzman, "Aerodynamic Drag Reduction Tests on a Full-Scale Tractor-Trailer Combination and a Representative Box-Shaped Ground Vehicle," *Society of Automotive Engineers* 750703, 5.

Many conference attendees dismissed the reliability of the coast-down method and questioned whether driving a truck in its actual environment could be cheaper or produce more accurate data than putting models in a wind tunnel could. It has to be said that Saltzman and his team had decades of experience in full-scale vehicle flight-testing and data reduction, and some knowledge of wind-tunnel testing, and knew well the benefits and limitations of each. It seemed that trucking industry representatives at the conference had only limited experience with the former, and were drawn almost exclusively to the idea of small-model wind-tunnel testing. They found inconceivable the idea that full-scale testing was cheaper. "But the facts are that it was cheaper to do the real thing," Saltzman said. And "the real thing gave you the real results."[11]

Wind tunnels do allow more control over variables but results obtained in them are not always accurate, and the variables do not always match real environments. The University of Maryland's Trailmobile tests of the early 1950s, for example, generated results similar to those of Saltzman's group with its Shoebox experiments, and the Trailmobile tests were conducted entirely on models in wind tunnels. General Motors also used wind tunnels to examine vehicular drag with an eye toward its own trucks, and released the results in 1961. While the similarities in the results of these three tests are undeniable, extrapolations based only on models and wind tunnels involve a certain leap of faith. There is even now an abiding notion that computational fluid dynamics (CFD) is the preferred method when trying to anticipate experimental aerodynamic results. But, as Saltzman noted, "anybody who's worked on full-scale vehicles—whether they be ground vehicles or aircraft—realizes you don't get the real answers that way—*yet*. They'll get some right answers, but they're laying themselves open to being fooled."[12] This said, even Saltzman agrees that carefully conducted wind-tunnel tests can now closely match results derived through full-scale tests, as both he and other researchers demonstrated over the course of the studies.[13]

The reaction within the trucking industry was not dramatically different from what the team periodically encountered at the FRC, but not because their colleagues doubted the tools they used or the results they were getting; many questioned the appropriateness of a NASA aeronautics center conducting research on road vehicles. More than a few at the center considered it an unnecessary diversion of resources. Between 1973 and 1975, the year in which the long-haul truck fairing research began, the center's budget increased from $11.7 to $13.2 million.[14] Given the overall reduction in NASA's budget, any increase

---

[11] Edwin J. Saltzman interview with author, Bakersfield, CA, 26 April 2005. For a larger discussion on wind-tunnel data versus the results of airborne experiments, see Milton O. Thompson, with a background section by J. D. Hunley, *Flight Research: Problems Encountered and What They Should Teach Us* (NASA SP 2000-4522, 2000). See also Vincent U. Muirhead, *Final Report on An Investigation of Drag Reduction on Box-Shaped Ground Vehicles* (Lawrence: The University of Kansas, 1976), KU-FRL, 180. Muirhead and a team of students conducted tests on a variety of 3/8-scale models that mimicked the full-scale test beds at the FRC; their results were within +/-3.3 percent of the FRC results.

[12] Saltzman interview with author. Saltzman has referred to this as one of the three "hurdles" he and his team had to contend with. These hurdles, public perceptions against which he and his team labored, were, in order of their appearance: the notion that testing of a four-foot model in a wind tunnel would cost less than driving a truck on the open road; the idea that computational fluid dynamics would be more accurate than testing the vehicle in its natural environment; and the notion that driving a heavily faired cab was tantamount to driving a "sissy truck," a comment the NASA driver endured at truck stops during road tests. On the matter of the continuing importance of wind tunnels in the face of a growing reliance on CFD, see Edward Goldstein, "Wind Tunnels: Don't Count Them Out," in *Aerospace American* April 2010, vol. 48, no. 4, 38-43, and Philip S. Anton, Eugene C. Gritton, Richard Mesic, Paul Steinberg, Dana J. Johnson, Michael Block, Michael Scott Brown, Jeffrey A. Drezner, James Dryden, Thomas Hamilton, Thor Hogan, Deborah Peetz, Raj Raman, Joe Strong, William P. G. Trimble, *Wind Tunnel and Propulsion Test Facilities: An Assessment of NASA's Capabilities to Serve National Needs* (Santa Monica: RAND Corporation, 2004). Both sources argue that while CFD has made great strides, wind tunnels are invaluable in validating CFD predictions and reducing costs by testing models in ways for which CFD is no substitute before moving to full-scale models, to name but two reasons for their continued use alongside computer simulation. The tension between wind-tunnel advocates and CFD advocates continues unabated, however. Inevitably, designers of aircraft first fly the vehicles to validate wind-tunnel and CFD work before delivering them to the customer.

[13] "It is this real life condition that often gets wind tunnel researchers in trouble in correlating between wind tunnel results and road testing results." N.C. Wiley, President, Airshield Division, Rudkin-Wiley Corporation, "Demonstration Aerodynamic Drag Reduction for the purpose of Reducing Fuel Consumption of Trucks," Stanford, CT, 5 December 1974, 13.

was a victory, but few at the center felt comfortable, and not without cause. The center's budget had fallen in 1966 and did not match, let alone exceed, its 1965 funding level (uncorrected for inflation) until 1971. The space program had taken a progressively larger share of NASA's budget throughout the 1960s, and the nation's economic circumstances by the 1970s were such that, even without the space program's drain, the agency's budget was not going to be as large as it had been. As historian Michael Gorn put it: "Aeronautics expenditures fell under headquarters scrutiny after James Webb's successor, Thomas O. Paine, resigned in September 1970. [James] Fletcher accepted cost cutting as a necessary measure." [15]

The X-15's visibility to the public and the fact that it consumed an exceptional amount of the center's budget meant that when it was cancelled (1968) it "raised questions about the survival of the FRC itself, a suggestion heard in such high places as the Senate Appropriations Committee."[16] Efforts by FRC director De Elroy Beeler to secure the center's position by reorganizing the agency's aeronautic centers along more rational lines met resistance at headquarters and among other NASA aeronautic centers; had those efforts been acted upon, by 1973 the FRC position may have meant reduced criticism of the truck fairing research program. There remained the fairly obvious fact that truck fairing research was not directly linked to aeronautics, however.

And so, although the FRC continued to host the lifting bodies for awhile longer, the era of highly visible, high-performance experimental flying seemed to many at the center to be at an end. It should be recalled that even in the midst of the Apollo program, between 1966 and 1968, the agency underwent what were termed Reductions in Force that saw numerous Apollo program engineers forced to look for work elsewhere. Despite what apparent logic there may seem today in conducting aerodynamic research on trucks, enthusiasm at the time was not uniform within the FRC for the research Saltzman, Montoya, Steers, and others were conducting.

Saltzman, Montoya, and Steers, however, were undeterred by the criticisms of their methodology and results. Their experiences with both types of test methods was something those in the automotive and trucking industries did not yet possess, and access to supercomputers necessary for sound CFD simulations was not yet readily available. But more to the point, testing in real conditions rather than the limited conditions of a wind tunnel was and continues to be a valuable capability and critical to resolution of some design problems.

---

[14] Richard P. Hallion and Michael H. Gorn, *On the Frontier: Experimental Flight at NASA Dryden*, (Washington, D.C.: Smithsonian Institution, 2003), 360.

[15] Michael H. Gorn, *Expanding the Envelope: Flight Research at NACA and NASA* (Lexington: the University of Kentucky Press, 2001), 298-299.

[16] Ibid.

Chapter Five
Shifting the Paradigm

Following experiments aimed at improving a truck's aerodynamics through use of existing aftermarket products, NASA's team at the Flight Research Center decided to try its own hand at modifying a long-haul tractor-trailer unit, "smitten by the challenge of defining the potential for reducing the fuel consumption of ground vehicles."[1] Capitalizing on what they'd observed with the aftermarket products, as well as on logic and experience, the FRC engineers designed an aerodynamic fairing to test on the center's cab-over. From midway back on the roof of the cab rose a bump that swelled upward to become a fairing in front of the trailer. Unlike the aftermarket products, the structure spanned the cab's entire width—and, of course, that of the trailer's as well. Dubbed the "Bat Truck," the crew sent it out for the customary road tests. To their dismay, their design yielded little drag improvement. Between the end of 1974 and 1975 the engineers devised an extension of the fairing so that it sealed the gap between cab and trailer (drawing on their own data from the aftermarket tests and existing information), yet even that did not generate sufficient returns to excite them.

To an eye accustomed to seeing long-haul trucks of the twenty-first century, the original iteration of such a vehicle designed by the NASA engineers looks like a surprisingly halfway attempt. This seemingly tepid experiment s a first effort was because of an enduring characteristic of the agency, and particularly this

An earlier view of the "Bat Truck," as it became known, showing the extent of the work done to clean up the area between the cab and trailer. The gap seal has not yet been attached.
NASA   E74-28088

The Bat Truck reflected both existing designs and original thinking by Dryden engineers. The curve from the roof of the cab up to the trailer was an adaptation of available products, reflecting the belief that the trailer constituted the greatest aerodynamic problem. The center's design resembled aftermarket devices that attached to the cab's roof and were meant to deflect air in front of the trailer. The effort to seal the gap between cab and trailer was a first, however. No effort had yet been made to clean up the front of the cab.
NASA   E75-28231

---

[1] Edwin J. Saltzman, "A Summary of NASA Dryden's Truck Aerodynamic Research," 821284 *Society of Automotive Engineers* (Warrendale, PA, 1982), 2. This is the published version of the paper Saltzman delivered to the SAE "Truck and Bus Meeting and Exposition," held in Indianapolis, IN, 8-11 November 1982.

The Bat Truck, right, alongside an almost identical test truck that served as the baseline vehicle. Both tractor-trailer units were sent on multiple trips around a 312-mile loop of highways to evaluate modifications made to the Bat Truck.
NASA   E75-28234

center. Despite a long and stunning record of success with unusual and often very dangerous aircraft, NASA and its predecessor, the NACA, had built a reputation on steadfast commitment to careful and methodical advances in research, not huge, sudden leaps. It was this, among other things that gave the agency such an enviable record of safety, if it also frustrated those inclined to a faster research pace. Not surprisingly, then, the first in-house test vehicle manifested what seemed to be the best traits of the most successful aftermarket products they tested, rather than some entirely new design.[2] Undeterred by the results, the

This picture of a Kenworth T600, the first production tractor with aerodynamic fairings of real consequence, reveals a fairing arrangement echoing that of the Bat Truck.
NASA   IMG 5142

---

[2] For a broader discussion of NACA/NASA character traits as an agency, see Curtis Peebles, *The Forgotten X-Planes: Configuration Research Aircraft of the Supersonic Era* (NASA Monographs in Aerospace History), forthcoming 2011.

A Kenworth T600 pulling a refrigerated trailer. The chiller is mounted on the nose.
NASA   *ED10-0030-5*

engineers scrapped almost all the initial modifications that had been made to the tractor.[3]

At virtually the same time this was happening, the engineers began a series of tests with two other FRC vehicles, a two-axle truck and a station wagon. The truck had a cube-like box on the back, typical of the period, with all edges forming 90° angles. Seeking to verify the results achieved with the Shoebox, they began with a series of coast-down tests to establish the baseline drag data for the truck, followed by trips up and down the South Base runway with tufts of yarn attached to the cargo box to show airflow patterns. They then proceeded to attach several devices, one at a time, and conduct the usual coast-down tests in order to determine the benefits, if any. Finally, the engineers turned the truck over to the fabrication shop to have a radius applied to the leading edges of the box's corners, in much the same way they had done with the Shoebox. Once again, the team took the truck out

Center engineers tested the two-axle truck with a device called a "flow vane" attached to the front of the cargo box roof. They tried three configurations of the flow vane. Tufting is a good indicator of the device's benefit.
NASA   *E75-29089*

---

[3] Ironically, others adopted the same design approach—without seeing the Bat Truck—and that shape remains on the road as of this writing despite its overall inefficiency.

This image reveals the gap at the top of the flow vane, which resembled a fixed leading-edge wing slat.
NASA   E75-28964

for a series of coast-down tests. The best results, a 30 percent reduction in aerodynamic drag, were realized with the truck when it had simply the cargo box sporting corners with radii.[4]

To the station wagon the team affixed a large valance similar to the one attached to the tractor-trailer used to carry the aftermarket fairings they'd tested for the DoT. And again, the front of the car was tufted so the airflow could be mapped. Both of these tests served to validate earlier work done with the Shoebox and with the tractor-trailer with the add-on fairings. This approach, seeking verification of earlier conclusions, was one to which engineers at the center were accustomed. It was no different than taking wind-tunnel predictions and validating them with a flying airplane,

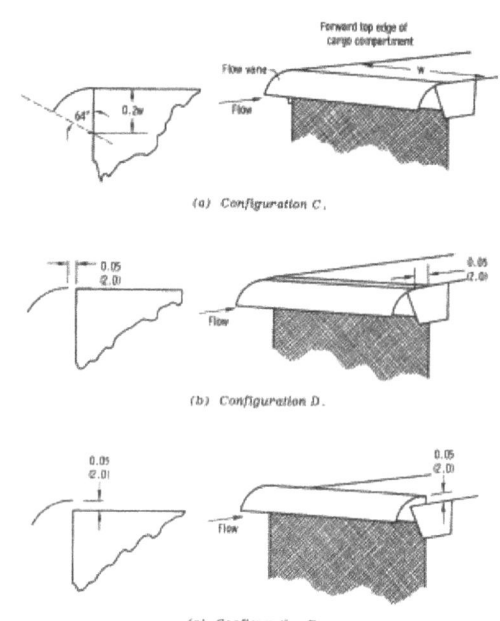

Fig. 5. Flow-vane configurations. Dimensions are in meters (inches).

Configuration	c_{D_a}	Aerodynamic drag reduction, percent	Total drag reduction, percent
A	0.875	---	---
B	0.610	30	26
C	0.808	8	7
D	0.815	7	6
E	0.794	9	8

Flow vane configurations applied to the standard truck at the FRC. Results can be matched to the table above and in figures 5 and 12.
NASA Technical Memorandum TM-72846

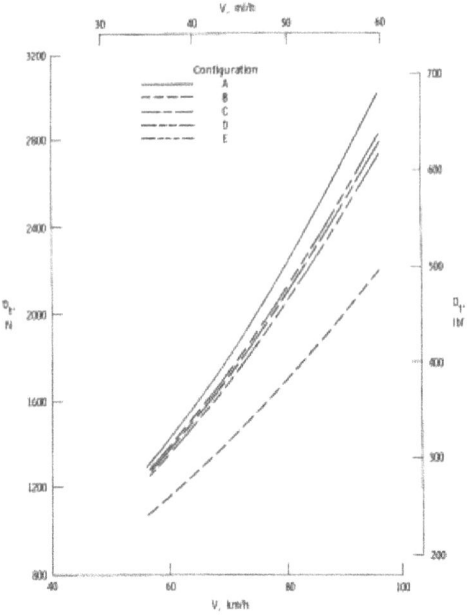

Fig. 12. Composite plot of total drag versus vehicle velocity for all configurations.

---

[4] Arthur E. Sheridan and Steve J. Grier, *Drag Reduction Obtained by Modifying a Standard Truck*, NASA TM 72846 (Edwards, CA: NASA Dryden Flight Research Center, 1978).

Two images of the two-axle Flight Research Center truck that was pressed into service for aerodynamic study. The image at left shows the truck with a stock cargo box. (For purposes of the study, engineers had the fabrication shop build a nearly two-foot extension of the cargo box, which was attached to the front of the box just behind the cab. It had the exact dimensions of the cargo box. This would allow engineers to modify the "front" of the box without permanently modifying a working vehicle.) Tufting maps a pattern of chaotic airflow along the side of the cargo box. The image at right shows the same truck, this time with the front vertical and horizontal edges sporting a radius. The tufting now aligns in a fairly uniform pattern. This change produced an aerodynamic drag reduction of 30 percent compared to that of the baseline truck.
NASA   *E74-27963, E74-27678*

or, more immediately, putting the reference flat plate on top of the delivery van to validate the preliminary tractive drag predictions the engineers had established at the project's outset.

Bat Truck Redux

Undeterred by the poor results of their first attempt at building a genuine aerodynamic tractor-trailer unit, the engineers began again, this time with a clean slate, choosing not to think in terms of small, aftermarket products but, instead, of how much improvement might reasonably be expected with cab-over trucks and trailers. They had in mind fundamentally redesigning the tractor, and brought to bear the full extent of their experience with the Shoebox as well as the knowledge of what had *not* worked in their previous experiment. Their effort led to the most radical tractor-trailer unit of the period.

  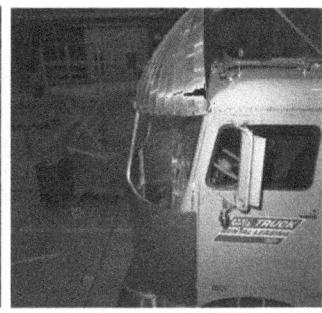

Poor results with the Bat Truck led center engineers to try again, this time incorporating as much as possible of all data acquired with the Shoebox and the DoT tests. Keeping the Bat Truck's gap seal between cab and trailer, they had the fabrication shop apply a new fairing to the cab (left). Highlighted in these three images are the sweeping curve of the fairing rising from the front roof of the cab to the height of the trailer (center), the radius applied to the roof fairing and the front corners of the cab, and the radius and deep valance at the front of the cab (right). More important, the collection of images underscores how dramatically these changes departed from the standard tractor design of the day.
NASA   *E75-28746, E75-28747, E75-28749*

 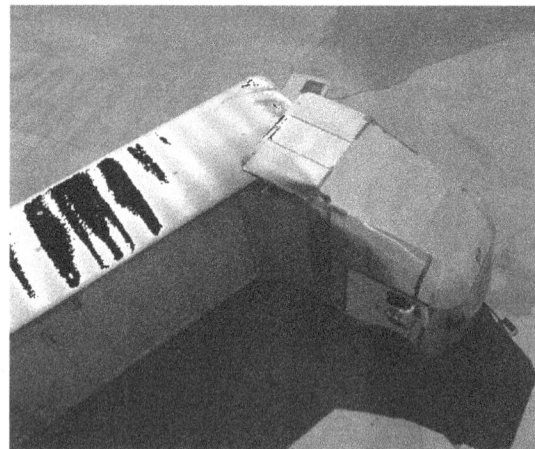

These two images show the newly faired tractor attached to the trailer. The first reveals two notable items: the dramatic change from an abrupt and aerodynamically "dirty" cab on the Bat Truck (left), and the completely enclosed exhaust stack, a feature that yields real aerodynamic benefits. The second image (right) illustrates a NASA first, the completely sealed cab-to-trailer gap. The side panels were hinged and spring loaded to permit them to flex as the truck turned corners, a feature that first appeared in the Bat Truck.
NASA
E76-29278, E76-29957

In much the same way as it had with the Shoebox van, the FRC sheet metal shop attached a framework to the existing tractor cab, onto which sheets of aluminum were fixed. These sheets were curved, as they had been in the most efficient version of the Shoebox, with a radii equal to 20 percent of vehicle width. The fairing started at the front axle, curved up and over the cab, and reached the same height as the trailer. It also ran along the side of the cab, closing the gap between tractor and trailer entirely. But the fairing did more than seal off the gap. The fairing's sides and top extended over the trailer and were hinged at the cab; small rollers on the inside of the fairing allowed it to flex side to side as well as up and down as the truck turned left or right or rode over uneven pavement, while bungee cords drew the sides of the fairings together to create constant tension on them, keeping the fairing flush against the trailer.[5] This ensured that the seal remained intact at all times. Though a somewhat familiar shape today, the truck looked like nothing else on the road at the time.[6]

Between the summer of 1975 and the spring of 1976 this newly configured test vehicle was subjected to the same 312-mile loop as the previous aftermarket configurations had been.[7] Following the usual coast-down method of establishing base drag for the combination before and after modification, the engineers sent drivers out with the truck for fuel consumption tests. Again, they sent along an almost identical but unmodified tractor-trailer unit for comparison and drivers of both trucks stayed in communication to ensure common shifting and the like. For their efforts Saltzman's team earned a 37 percent decrease in aerodynamic drag over the unadulterated tractor-trailer at an average speed of

---

[5] The hinges and rollers were inexpensive items obtained from a local hardware store.

[6] Louis L. Steers and Edwin J. Saltzman, "Reduced Truck Fuel Consumption through Aerodynamic Design," *Journal of Energy* vol. 1, no. 5 (September-October 1977): 313. Some years earlier small models of faired COE tractor-trailers had been tested in wind tunnels by both the University of Maryland and General Motors, but neither tested a full-scale version on the road. See A. Wiley Sherwood, *University of Maryland Wind Tunnel Report No. 85: Wind Tunnel Test of Trailmobile Trailers* (University of Maryland Wind Tunnel Department: College Park, 1953), and Harold Flynn and Peter Kyropoulos (General Motors Corporation), "Truck Aerodynamics," presented at the SAE International Congress and Exposition of Automotive Engineering, January 1961 and published in *SAE Transactions* (1962) vol. 70: 297-308.

[7] As before, the truck was a White Freightliner cab-over-engine with a sleeper compartment. The trailer was a dual-axle, smooth-sidewall type, and the combination weighed 32,000 pounds.

This side view of the modified tractor-trailer combination, at right, emphasizes how nicely faired the tractor is as well as the changes to the front of the cab, especially when compared to either the Bat Truck or to its stock counterpart, parked alongside.
NASA   E76-29993

55 mph, and only slightly less at slower speeds.[8] The tests showed a reduction in total drag for the highly modified rig of 28 percent at 50 mph and above, yielding a reduction in necessary power at 55 mph and 65 mph, respectively, of 40 and 49 horsepower.[9] The FRC group's modifications translated into conservatively measured improvements in fuel consumption of "20% in light wind conditions and 24% in near-calm conditions while operating at 55 mph."[10] To be sure, wrote Steers and Saltzman, these figures would vary with higher winds and stronger crosswinds.[11]

Pleased with their success, the engineers set about

Front views of the highly faired tractor, showing the smooth curvature of the cab's front as well as the method of attaching sections of the fixture to make the structure accessible and removable. Note that the radiator door is closed in the first image (left), and a skunk has been coyly affixed to the truck's top (right), homage to Lockheed's well-known secret research unit, the Skunk Works.
NASA   E76-29967, E76-29958

---

[8] Louis L. Steers and Edwin J. Saltzman, "Reduced Truck Fuel Consumption Through Aerodynamic Design," *Journal of Energy* vol. 1, no. 1 (September-October 1977): 316.

[9] Ibid., 315.

[10] Ibid., 316.

[11] Steers and Saltzman, "Reduced Truck Fuel Consumption through Aerodynamic Design," *Journal of Energy*, 316; and R. A. Servais, "An Experimental and Analytical Investigation of Truck Aerodynamics," *Proceedings of the Conference/Workshop on Reduction of Aerodynamic Drag of Trucks* (Washington, D.C.: National Science Foundation, October 1974), 55-61.

writing formal reports of their research, both for their peers and for a larger audience. The center also undertook promotional work, sending out news releases about the truck and the results. This was, after all, what NASA does with its research.

Meanwhile, in late 1977 a young graduate student from the California Polytechnic University, San Luis Obispo, joined the FRC truck research team. Randall Petersen was assigned to the Shoebox test vehicle to conduct a boattail study. His work was to be part of his master's thesis. Tufting studies showed that when the vehicle's front corners were rounded, the airflow remained attached to the side of the van, something considered desirable for aerodynamicists since this means the flow is smooth and is not generating drag-producing vortices. Knowing a thing or two about the shape of aerodynamically efficient (and inefficient) vehicles, the group had the fabrication shop craft and attach a boattail to the Shoebox. This is a shape that tapers nearly to a point, drawing together the vehicle's four slightly curved surfaces to a negligible cross-sectional area. The point was to help coax the air to merge at the aft end of the van and eliminate the van's low base pressure. To the FRC engineers' satisfaction, the tufting research guided by Petersen showed the air following the shape of the boattail, for the most part. Unexpectedly, the tests showed that the airflow remained attached up to a point well along the fairing, after which it separated. This meant they could cut off the boattail at the point of separation without sacrific-

These two images show the Shoebox in its first and second phases. At top, the vehicle's edges meet at 90 degrees. Below, the four vertical corners have been given a radius. The effects of this are evident in the tufts of yarn attached to the Shoebox's side. Top, they are splayed in all directions, indicating chaotic airflow; bottom, they flow fairly uniformly.
NASA   E74-26992

Three images of the now-completely modified tractor unit, prior to testing. The image at left, taken from the rear, shows the extent of the fairing above the cab as well as the manner in which the fairing extends over the gap seal between cab and trailer. That gap seal is attached to the tractor's frame, not the trailer. The middle image shows the new valance in its lowered position while also illustrating the difference between it and the old bumper. The image at right reveals just how completely rounded the new cab is when compared to the old structure, visible beneath the aluminum and Plexiglas, and shows the lowered valance.
NASA   E76-29265, E76-29267, E76-29268

The Shoebox, extensively modified fore and aft. Once a vehicle with flat sides and 90-degree corners, it now had both a front and a back that were dramatically curved. Visible behind some of the aluminum is the original van. The lever at the front connects to a shaft that reaches the driver and enables him to open the door to provide air to the radiator.
NASA
E78-33683

ing significant drag reduction, shortening the overall length. The boattail reduced the vehicle's aerodynamic drag another 32 percent, even in its truncated form.

The FRC research results were extended when researchers at the University of Kansas conducted additional tests using that institution's wind tunnel and a fairly accurate 1/25-scale plastic model of the tractor-trailer unit used by Saltzman's team. (This followed research at the university's wind tunnel with models of the center's modified Shoebox to "improve air flow over the front top [and] bottom and investigate means of reducing the base pressure" beyond

In its final configuration, the Shoebox had a boattail attached to it. Testing proved that having the entire boattail was unnecessary since airflow began to separate from its surface before reaching the tip. These two images show the full boattail as well as the point of separation (right), and the truncated boattail (left).
NASA

Around the same time that center engineers tested aftermarket add-on devices for the DoT (c. 1974), they decided to try some modifications to the center's station wagon (left). The effort was aimed at controlling the airflow at the front of the car, thereby influencing its progression around the car. The blunt front that extends below the standard bumper and valance is common on modern automobiles, but it had only begun to appear on racecars of the era, and even then such a valance did not always span the width of the car. A similar arrangement appears on the truck used in the DoT tests (right).
NASA
E74-27885, EC74-4214

that which the NASA engineers had achieved.[12]) The University of Kansas group, led by professor Vincent Muirhead and working under a contract from NASA, wanted to examine both the effects of side winds on the tractor-trailer units in a controlled environment and derivative modeling. They started by establishing baseline drag data for the model, and proceeded to add fairings of different kinds, testing each new configuration in the wind tunnel. Having reproduced the NASA configuration satisfactorily, the Kansas group then applied its own experimental fairings. They added side panels between the bottom of the trailer and the road, ultimately encasing the trailer's underbody and closing that gap. They even went so far as to cover the rear wheels of the trailer with side panels. In its final configuration they applied a rounded boattail to the back of the trailer, much as had the NASA engineers on the Shoebox. For its most aerodynamically faired configuration, their experiments showed an aerodynamic drag reduction just shy of 60 percent in a windless environment, 67 percent when a side wind blew on the unit.[13] This figure, with its attendant reduction in fuel consumption, was startling. Based on the drag reductions of the FRC's highly faired tractor-trailer,

---

[12] Vincent U. Muirhead, *An Investigation of Drag Reduction on Box-Shaped Ground Vehicles* (Lawrence: University of Kansas Center for Research, Inc., 1976), 1, 4 (also issued as NASA CR-163111).

[13] Sheridan and Grier, *Drag Reduction Obtained by Modifying a Standard Truck*, and Randall L. Petersen, *Drag Reduction Obtained by the Addition of a Boattail to Box Shaped Vehicle* (Edwards, CA: NASA CR 16113, 1981), 1, 5, and Vincent U. Muirhead, *An Investigation of Drag Reduction for Tractor-Trailer Vehicles*, Edwards, CA: NASA CR 144877, October 1978.

expressed as A, and the most highly faired model by the Kansas group (the model with the boattail and under-trailer fairing), expressed as B, and assuming an average annual mileage of 100,000 driven by an independent trucker, the savings were calculated as 3,435 gallons and 6,829 gallons of diesel fuel per year for A and B, respectively.[14] In 1979 Saltzman himself noted that if just the tractor-trailer modifications applied to Dryden's full-scale test vehicle were adopted by the nation's trucking industry, it could save 26.3 million barrels of oil a year.[15]

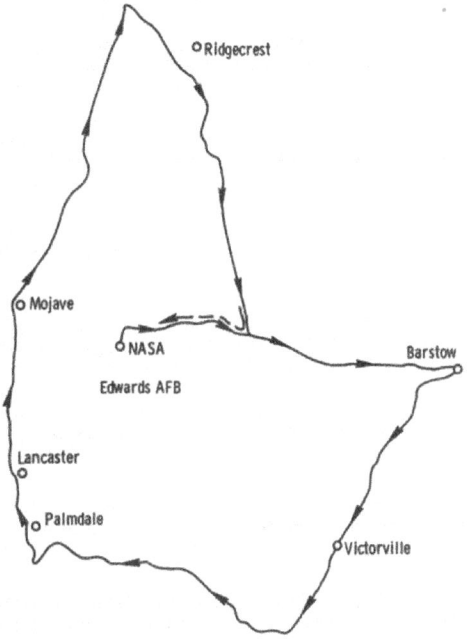

The NASA fuel-consumption test route driven with the pair of tractor-trailers. The route was primarily over open roads, including Highway 58, but some of the route took the trucks through urban areas as well.

---

[14] Vincent U. Muirhead and Edwin J. Saltzman, "Reduction of Aerodynamic Drag and Fuel Consumption for tractor Trailer Vehicles," *Journal of Energy* vol. 3, no. 5 (September-October 1979): 283-284.

[15] Saltzman's comments are attached to NASA Technology Notes, "Low Drag Truck Design Tested by NASA," NASA Ames Research Center (n.d.). At the time the Flight Research Center had been incorporated into the Ames Research Center, hence the release came from the ARC.

Chapter Six
Technology Transfer

The continued economic and emotional pressure of the gas crisis fed interest in the NASA research, and it took remarkably little time for news of Dryden's results to begin filtering beyond the professional journals and into the popular media. (In 1976 the Flight Research Center was renamed the Hugh L. Dryden Flight Research Center, in honor of Hugh Dryden, director of the NACA from 1947 until the creation of NASA in 1958, and perhaps the most significant force behind the realization of the X-15.)

In April 1975 *Car and Driver* magazine published "Project Aerodynamic Van: Trying to Turn a Shoebox into a Slipper," an article recounting the writer's efforts at improving the aerodynamics of a typical passenger van. Reflecting the pressing issue of the day—the price of gas—Don Sherman began the article with thin wit: "For all of us who can't list King Faisal as a personal friend, fuel economy is a mounting concern."[1] From quacks to serious interested parties, he added, everyone was attacking the problem after "a flurry of recent activity in this field extending from NASA research." For their part, Sherman and a crew from the magazine decided to test a host of devices and theories, from popular ideas to university data. They selected a standard passenger van and began their tests.

> Our fondest hopes were for the most involved alterations we attempted. Wind tunnel testing at the University of Utah had demonstrated tremendous reduction in drag (up to 45 percent) on large semi-trailer trucks by adding a horizontal roof duct to the trailer. The idea is to catch high-pressure air flowing upward across the windshield of the tractor (or the leading edge of the trailer) and to duct it rearward and empty it into the low-pressure area created in the wake of any box-like object moving through the air.[2]

So Sherman and his team built a duct from plywood and aluminum sheeting that snagged air ahead of the windshield and ran it along the roof, then turned it downward 45° at the back of the van. It weighed 50 lbs. "Instead of drag reduction, we found the roof duct created huge penalties. Compared to the base van, it added about 39 percent in aerodynamic drag—almost as if the University of Utah's wind tunnel had misplaced a minus sign." Sherman did admit to the likelihood, however, that the crew's own duct design and construction were more at fault than the idea itself.

Their next modification was a smooth belly pan attached to the underside of the front of the van. In adding the pan, Sherman noted that NASA experimenters had identified a gain of 15 percent with this technique. *Car and Driver*'s results didn't match the work by the NASA researchers, but the benefit of "a smooth underside was obvious," wrote Sherman. Perhaps their work wasn't as meticulous as the effort at the FRC. Next on the list of changes was an aftermarket nose spoiler (made by Karvan) mounted to the front of the van: using it, they found a disappointing drag *increase* of 3 percent.

The modification that achieved the best results was discovered through one of their own experiments. They removed the van's two outside mirrors and found that doing so yielded a 6 percent drop in aerodynamic drag—more than from any other modification attempted. "Did we solve the problem?" asked Sherman. "Frankly, no—but our testing did reveal several seemingly good ideas that simply *don't* work, as well as a few areas that *can* deliver a genuine boost in highway fuel economy."[3] Without doubt the less-than-scientific nature of these tests influenced the results, but the effort was more than whimsy. Despite a tongue-in-cheek writing style, Sherman and his group were trying to address a serious issue while drawing popular attention to the fact that altering a

---

[1] Don Sherman, "Project Aerodynamic Van: Trying to Turn a Shoebox into a Slipper," *Car and Driver* April 1975, 57.

[2] Ibid., 60.

[3] Ibid., 57, 60, 100.

vehicle's aerodynamics could indeed have an effect on the pocketbook.[4]

Coming even closer on the heels of the work at the FRC was a study done by Peter Lissaman of the firm AeroVironment, Inc., in California. Lissaman and a team began work in 1974 and published their results in a paper given in 1975 titled "Development of Devices to Reduce the Aerodynamic Resistance of Trucks." "It is a matter of interest," began Lissaman, "that the aerodynamic drag of road vehicles consumes about 7% of the *entire* energy consumed in the United States."[5] Little things counted when it came to aerodynamic drag, as the unschooled Sherman and his group at *Car and Driver* discovered. Simply removing the rain gutters from around the doors of a Volkswagen Karman Ghia could reduce the car's aerodynamic drag, noted Lissaman.[6]

In 1974 AeroVironment took a tractor-trailer assembly to El Mirage Dry Lake in California's High Desert region and, with the permission of the Bureau of Land Management, drove the truck on the lakebed, an almost perfectly flat surface. AeroVironment used funding from the National Science Foundation to conduct the tests. Engineers applied a full fairing over the cab of the truck, making it look very much like the assembly that FRC technicians applied to their tractor-trailer unit except that the front showed no radiused metal work. The gap between cab and trailer was sealed in much the same way as it had been during the FRC tests. AeroVironment test results identified a variation in results between 2 and 5 percent despite best efforts at continuity in conditions; even though their truck was carefully and extensively instrumented, drawing conclusions was difficult. (It carried an HP 9820B computer in the cab to manage incoming data.) To remedy the problem, the AeroVironment team then drove the truck three times on a 90-mile road course to subject their modifications to normal driving conditions. They found that their faired cab and trailer yielded an aerodynamic drag reduction of approximately 23 percent, for a fuel savings of 18 percent–not dissimilar from FRC results. AeroVironment had a variety of results with the aftermarket devices they tested ranging from 8 to 20 percent reduction in aerodynamic drag between configurations–much as the NASA team at the FRC found in its earlier investigation.[7]

Interested in what could be done with a smaller vehicle, AeroVironment attached a "lip" to one single-axle truck and tested it against another lacking the modifications. Again, they carefully instrumented the trucks and even placed meters on the fuel systems to measure flow rates as well as engine rpm, then drove the two trucks in crosswinds of up to 20 mph. From these experiments they learned two things of particular interest. In heavy crosswind the truck with the lip maintained a 55 mph speed while the one without any modification was unable even to reach 55 mph. And they saw a drop in aerodynamic drag of about 34 percent as well as a fuel mileage gain of 23 percent for the modified truck.[8]

With the road testing done, AeroVironment then conducted wind-tunnel tests at the California Institute of Technology (Cal Tech) to validate the modeling worked up on the road. They put into the wind tunnel 1/24-scale truck models "identical to the full-scale vehicle used in the desert and road tests" only to find that "the results of the wind tunnel tests did not correlate well with those obtained in direct field testing." But as with the NASA team at the FRC, AeroVironment found that the gap between cab and trailer was a major source of disturbed air and, consequently, drag.[9] (Increasing the space between tractor and trailer

---

[4] The article did not say how the team established a baseline for the van's efficiency, but removing the exterior mirrors was, in fact, a tactic others also tried with real benefit.

[5] P. B. S. Lissaman, "Development of Devices to Reduce the Aerodynamic Resistance of Trucks," Society of Automotive Engineers annual meeting, Seattle, WA, 11-14 August 1975, 1. At a 1974 meeting held on the Caltech campus in Pasadena, CA, AeroVironment presented results of its research to that point. Schenck, "New Focus on Air Drag," *Trailer/Body Builders*, November 1974, 36-41.

[6] There is no indication that Sherman knew of Lissaman's research or conclusions about mirrors. The article refers to "removing weather stripping" but a Dryden engineer familiar with the experiment and with the Lissaman work assures the author it was rain gutters.

[7] Lissaman, "Development of Devices to Reduce the Aerodynamic Resistance of Trucks," 5.

[8] Ibid.

[9] Ibid. Saltzman expressed his own reservations about the value of wind tunnels in cases such as this one, as well as his skepticism over computational fluid dynamics modeling, since neither adequately mimics real-world conditions. Saltzman interview with author, 26 April 2005.

increases air resistance. Conversely, narrowing the gap from 45 to 25 inches can improve fuel economy by as much as 2 percent beyond any gains already achieved with add-on aerodynamic devices.[10])

AeroVironment added a lip to the top front of a semi's trailer and after driving it found an aerodynamic drag reduction of roughly 9 percent and fuel savings of 4 percent. But they fared better when putting the lip on the top of the single-axle truck; in that case, they earned a 34 percent reduction in aerodynamic drag and a 23 percent improvement in fuel mileage.[11]

For the shield on the tractor unit, AeroVironment engineers opted to devise a screen rather than affix a solid shape. Their logic was that yaw effects on the trailer created by side winds, when added to the air pressure differential between the top of the cab and the trailer and the nearly flush sides of the tractor and trailer, combined to make flow separation a real problem. A plate atop the cab, therefore, full of round holes and with vanes on either side, would allow some air to enter the area just behind the cab and raise the pressure enough to counter the typical low-pressure zone in that area, which was susceptible to the buffeting of yaw effects on the tractor and trailer. The researchers believed that these perforations would largely negate the problems afflicting the truck and trailer assembly. They tested a model in a 10-foot wind tunnel and found an 11 percent reduction in drag, although Lisamann quickly reminded his readers of the risks in extrapolating too much from wind-tunnel tests. In calm winds, they found a reduction of 30 percent in aerodynamic drag during rolling tests, their equivalent of the coast-down method. The wind was a significant factor, Lisaman admitted, noting that on another night, with a crosswind of about 3 mph, the aerodynamic drag reduction fell. When compared to the flat plate type of deflector, they found the porous version superior, not to mention quieter because of the reduction in "cab roof" banging that came from wind buffeting.[12]

Industry Response

Based on the initial conference presentation made by the FRC team, it was apparent even by then that reducing the gap between cab and trailer was important (filling the gap has the same effect). This sentiment was reflected in the comments of others in the field as well. "The gap is especially important in crosswind situations, since flow through the gap can totally wipe out gains that are made in headwinds [fairing of the forebody]." The afterbody will be ignored, wrote journalist Paul Schenck, since it needed to be functional "so the attention is being focused farther forward."[13]

Not surprisingly, popular interest in improving the efficiency of long-haul vehicles spawned other attempts similar to those written about by *Car and Driver* magazine, some amateurish, some grounded in experiments. Even before NASA engineers had finished modifying their tractor-trailer unit there were new aftermarket products garnering attention. Some were familiar, such as the Airshield. Others were imaginative, if a bit odd.

A device dubbed the Batmobile because it unfolded accordion-like to fill the gap between cab and trailer (with fillets reminiscent of the Batmobile's fins), able to flex as the cab made turns, was one example. Drivers using the system reported fuel savings of 6 to 8 percent but it was noted that, "these savings would have been considerably more if the drivers had not tried to emulate Batman's speed. When an underpowered rig gets a boost from a drag reduction device, the drivers immediately try to increase speed which wipes out the fuel savings."[14]

---

[10] Laura Crackel, "Stretching The Limits," *Overdrive*, http://www.etrucker.com/apps/news/article.asp?id=39038 (accessed 3 June 2009).

[11] The FRC engineers had tried an aftermarket product similar to this one, a curved device that ran the width of the trailer but sat forward of it by 6 inches and left a 1.5-inch gap between it and the trailer top. See Device D in: Louis L. Steers, Lawrence C. Montoya, and Edwin J. Saltzman, "Aerodynamic Drag Reduction Tests on a Full-Scale Tractor-Trailer Combination and a Representative Box-Shaped Ground Vehicle," paper presented at the Society of Automotive Engineers annual meeting, Seattle, WA., 11-14 August 1975, SAE Publication No. 750703, p. 7. The FRC test results showed only 2-3 percent drag reduction with this device.

[12] Lissaman, "Development of Devices to Reduce the Aerodynamic Resistance of Trucks," 8.

[13] Schenck, "New Focus on Air Drag," *Trailer/Body Builders*, 36-41. Saltzman felt that all the experimenters of the period skirted work on reducing base drag (afterbody) because of the trucking industry's widespread use of aft doors to accommodate warehouse loading docks. Saltzman to Gelzer, notes on manuscript draft.

[14] Schenck, "New Focus on Air Drag," *Trailer/Body Builders*, 38.

The Aerospan Corporation, meanwhile, offered a "tent-like creation of heavy-duty reinforced fabric on a frame" that mounted to the rain gutter rails of the truck's cab. It looked like "a shapeless tarpaulin," the product advertisement acknowledged, until the truck built up speed. Then, a duct at the front of the contraption forced air into the fabric bag and presto, an inflatable air dam opened up. The company claimed a 10 percent fuel savings with the device and added that its collapsible nature meant it worked well in crosswinds.[15] Aerospan Corporation calculated that their device would appeal to all those drivers whose trucks had been manufactured too early for factory aerodynamic modifications but who still wanted the benefits of such features.

Overdrive, the self-styled "vehicle of the American trucker," catered to the independent operator. One of its regular features was a "dissection" article, included, according to the publisher, to verify the truth about a product. In October 1974 the magazine ran a story about add-on devices for trucks titled "Overdrive Foils Aerodynamic Wind Deflector."[16]

In that spirit the magazine asked for a sample of Camper-Flow Windbreaker's air dam before the magazine would accept its advertisement, and then tested two of the company's cab-mounted shields, noting ease (or difficulty) of installation and effectiveness. The magazine excoriated the company for sloppy market research, nearly impossible guarantee terms, and a poor commodity. It neither confirmed nor refuted, however, the product's alleged aerodynamic improvements.

The product was simply a triangular structure, flat on all sides, that the owner attached to the roof of the cab using the rain gutters, clasps, and suction cups. "It may well result in a 7% fuel saving, but *Overdrive* just cannot imagine an independent trucker spending $30,000 for a new tractor with sharp paint design and chrome options, and then attach this 'roof rack' type of device to that truck."[17] Although the product might well improve mileage, the design was inconvenient, to say the least (it was all but impossible to run a roof-mounted air conditioner on the COE to which they fitted the shield), and flimsy in its attachment.

Given what NASA and other researchers found out about air dams, it is questionable whether the Windbreaker would have been very effective, and equally questionable whether anyone would have been able to claim a refund if it wasn't. But it is a measure of the burgeoning interest in devices to improve the efficiency of long-haul trucks that a magazine would, as early as 1974, run such an investigative article in the hopes of foiling con artists.

In early 1975 the White-Freightliner company decided to test a collection of potential aerodynamic improvements to trucks in an operating environment. The company sent two rigs out on a 6,000-mile run from Indianapolis, Indiana, to Portland, Oregon, then down to Chino, California, back to Portland, and home to Indianapolis. The trucks pulled identical trailers loaded with the same weight. Each truck had a crew of two drivers that changed rigs every eight hours to reduce the effects that individual habits might have on the outcome. While they rode in proximity to one another, they stayed far enough apart to avoid drafting, and changed places every 100 miles in a further effort to balance the test. They even filled the trucks from the same pumps when they fueled to make sure their intake was at least equally metered.

The lone difference between the two tractors was in their setup. One had no modifications at all; the other, called the Energysaver rig, benefited from radial tires, a slightly improved engine that ran 200 rpm lower than the stock tractor while making the same speed, a Thermatic fan that engaged or disengaged depending on engine temperature instead of running constantly, and a Rudkin-Wylie Airshield and Vortex Stabilizer on the cab and trailer nose.

In addition to helping drivers determine that they preferred the driving characteristics of the modified truck ("It handled much better than the standard rig"), the experiment yielded a mileage improvement of 25 percent over the 6,000-mile course, a remarkable figure.[18]

---

[15] Ibid.,123.

[16] "Overdrive Foils Aerodynamic Wind Deflector," *Overdrive,* October 1974, vol. 14, no. 10, 88-91.

[17] Ibid., 91.

[18] "The New Mileage Misers," 3. This was a brochure from *HDT Heavy Duty Trucking* (March 1975) that originally appeared as an article in the same magazine. The International Harvester truck company conducted its own tests with a company rig and claimed benefits of 25 to 30 percent fuel savings, savings considerably greater than those of their competitor inasmuch as IH employed the same modifications, right down to the Airshield.

The specifics for the course were as follows:

Truck	Standard	Modified
Fuel used	1,481 gal	1,186 gal
Fuel mileage	4.45 mpg	5.56 mpg

At $.46 a gallon, the company expected a driver would save as much as $2,071 annually if he drove 100,000 miles per year—more than enough to pay for the aftermarket modifications in the first year (estimated to cost about $1,880 retail).[19] A story in *Family Safety* magazine in the summer of 1978 recounted a test conducted by United Parcel Service with a pair of trucks that made identical trips on a loop around Columbus, Ohio, one at 65 mph, the other at 55 mph. The slower truck used roughly 32 percent less fuel. By this time the national speed limit was 55 mph, established by the Nixon administration in 1974, and the test merely underscored the reason for the lower speed limit.[20]

The Ryder Truck System built ten prototype tractors, mindful of the growing fuel crisis. The units were still slab-sided, but instead of large sections of the cab positioned perpendicular to the motion of the truck, the cab's large, flat facets were angled to better suggest a wedge-like object going through the air.[21] The firm did not adopt the design, however, and the trucks eventually disappeared from the road.

Still, reaction among truck manufacturers to these developments was gradual. In part this is because retooling a production line, as well as completing the engineering work for a new vehicle, takes considerable time. In the case of the Kenworth Truck Company, what became the new and aerodynamic truck evolved in fits and starts, and it was nearly ten years before its first commercially available, stock aerodynamic truck reached the market. That truck marked the turning point in the industry regarding attention to efficiency and aerodynamics. Kenworth introduced the T600 in 1985, and the truck borrowed from NASA's research even while the company conducted its own wind-tunnel studies. The T600 was radical by the standard of the day, another reason trucks did not suddenly sprout new shapes in 1975. Moreover, the T600 did not incorporate all the modifications NASA had demonstrated to be useful. As Larry Orr, chief engineer at Kenworth, recalled thinking, the new design was likely to meet resistance, so he started out conservatively. "We had other ideas which would have made the T600 even more radical-looking. But we didn't want to introduce too much too soon."[22]

Kenworth's T600 was indeed a radical-looking truck, but its efficiency was the selling point—a claimed 22 percent fuel mileage improvement over the W900B from which it derived. Among the first to commit to the new design was Contract Freighters, Inc. (CFI) of Joplin, Missouri, a fleet operator with some 500 trucks moving around the country in 1985. Others, too, found the new design's fuel efficiency compelling enough that, by the end of its first year on the market, the T600 constituted 40 percent of Kenworth truck sales.[23] The number is not surprising considering the design offered the chance to reduce an operator's largest annual expense—fuel—by 22 percent and the typical long-haul driver might well buy 25,000 gallons of diesel in a year.

The role that fleet operators played in acceptance of the new aerodynamic design cannot be overstated, for in spite of the logic behind the new design, drivers

---

[19] The $.46 per gallon cost of diesel fuel dates from a 1975 publication. For perspective, one has only to consider the potential savings associated with an improvement of 1.11 miles per gallon in the context of current fuel prices. That figure would not incorporate improvements made since 1975, of course.

[20] *Family Safety Magazine*, summer 1978, vol. 37, no. 2, 2.

[21] "The Shape of Trucks to Come," *Trailer/Body Builders*. The first tractor was designed and built by Dean Hobbensiefken, himself a truck driver. He eventually sold the truck to the Ryder Corporation, which commissioned a firm to build ten more that were used in over-the-road tests. Though the new tractor had between 10 and 30 percent less drag than comparative tractors (depending on wind direction), the improvement wasn't entirely due to the new shape. Relocated radiators had cooling fans that ran only on demand and the truck itself ran on very different tires and wheels than did its competitors. The tractor also weighed less than the standard tractor of the period because of lighter components and an overall weight-reduction effort. None of this discredited Hobbensiefken's concept or the gains realized, many of which presaged those of later designs.

[22] "Half said it was the most unusual truck they ever saw; some just shook their heads," recalled Larry Orr. "Kenworth's T600 – A Look Back at the Truck that Broke the Mold," *Land Line Magazine*, 2005, 3.

[23] Ibid., 4.

were not especially happy about the new truck's shape. "I had some drivers come into my office literally in tears, threatening to quit if I made them drive what they called the 'anteater,'" recalled Glenn Brown, president of CFI. "They were ribbed in truck stops and on the CB" for driving the model. This was a similar reaction the driver of the original NASA cab-over had endured when taking the modified truck on its road test ten years before.[24] That same sentiment had surfaced even earlier, with General Motors' new Astro 95, a COE design from 1974 with rounded corners and a small air dam on the roof with compound curves (corners reminiscent of the NASA Shoebox design), but which was by no measure radical. "The GMC Astro 95 has a more streamlined shape than the boxy cabs of the prestige names," noted Schenck, "but many drivers prefer the tough-looking, brawny, hairy-chested box over the gentle curves of the Astro cab." Nevertheless, the purchasing power of fleet operators and their general indifference to their trucks' appearances and perceived manliness, compared with their interest in the bottom line, made them a key factor in the growing popularity of the new shapes of long-haul trucks.[25]

---

[24] Ibid., 5. "I recognized that it was innovative and different. We were looking for ways to improve our efficiencies and decided to order 100 of the trucks," said Brown. "We were hoping that the new design of the T600 was something we could use to get a head start on the rest of the industry. It was definitely a gamble to be the first in the market to put the truck on the road, but it proved out. Today, CFI has 1,650 Kenworth T600s in its fleet. What's more, the company is so fond of the T600 that it still has the very first T600 it purchased back in 1985." "CB" is a contraction of Citizen Band radio, a system of short-distance radio communication in the 27-MHz band with 40 channels and set aside specifically for use by the U.S. public. CBs became immensely popular even before the gas crisis began in 1973 because the federal government imposed a speed limit of 55 mph on national roads; CBs enabled drivers of trucks and cars to share information about speed traps, and featured prominently in a number of songs and motion pictures of the period.

[25] Schenck, "New Focus on Air Drag," 40.

Chapter Seven
Depressed Cows

A typical livestock trailer circa 2011.
NASA

IMG-5150

Not long after the release of the team's data in the form of NASA reports, articles, and papers given at professional conferences, Saltzman took a phone call from Dr. Floyd Horn, a Texas A & M University professor with extensive links to the U.S. Department of Agriculture. Interested in the Dryden engineers' results, Horn expressed concern about unanticipated consequences, reservations deriving from his own work with the livestock industry. The now-aerodynamic tractor, of which the engineers at Dryden were rightly proud, would, he suggested, pose a threat to livestock in transit since it promised to smooth the passage of air around the trailer. Horn feared this would reduce the flow of air into the trailer, with adverse effects on the cattle.

Horn's concern was not abstract: the loss to livestock farmers from "shipping fever" totaled between $400 and $500 million annually in 1980 dollars, and only slightly less in 1975, the year Horn approached Saltzman. Furthermore, the cost to livestock haulers and owners was not measured strictly in the number of deaths per trip.[1] Another factor was the effect the jour-

---

[1] Vincent U. Muirhead, *An Investigation of the Internal and External Aerodynamics of Cattle Trucks* (Edwards, CA: NASA CR 170400, 1983).

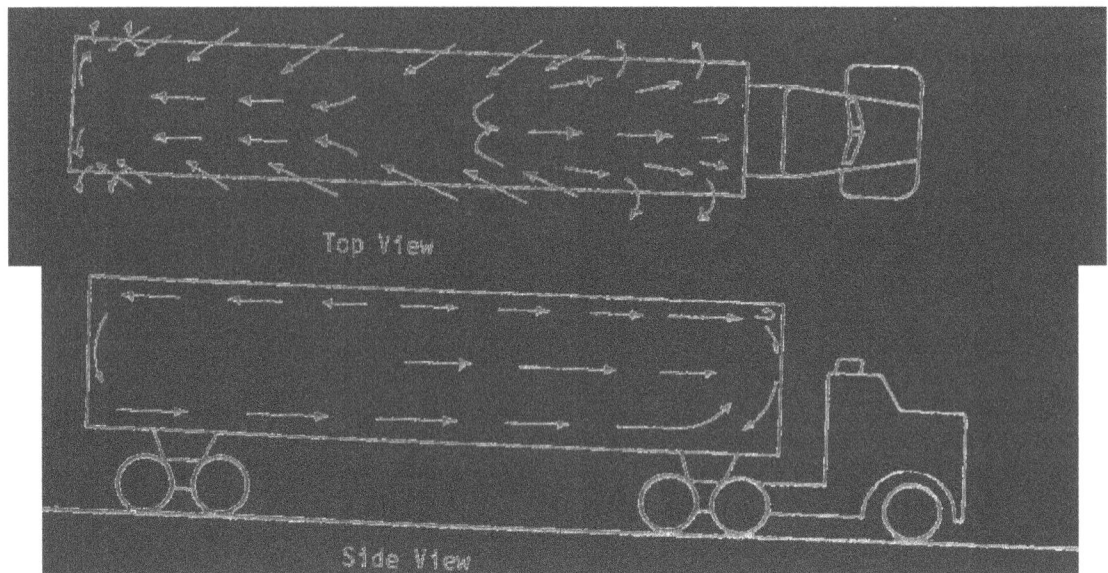

Airflow patterns in a typical livestock hauler, showing a great deal of recirculation and little inflow of fresh air. NASA patent

ney had upon the weight of the cattle on reaching their destination, as well as the quality of the meat; both suffered measurably. As late as 2005 the "shrinkage" rate varied from 1 to 8 percent for trucked livestock, something especially significant since weight is a central factor in the purchase price for cattle at abattoirs.[2]

A 1980 National Cattleman's Association report noted that "shipping fever" accounted for a 3 to 5 percent loss of stock during and immediately following transport.[3] The cause is complex but the contributing factors are clear enough: "overheating, uneven ventilation, [and] unfavorable air composition (dust and fumes due to ingestion of animal-generated moisture and ammonia vapor)," a pungent description indeed. This did not include the stress endured by the animals during the trip, added the report, or stress to mothers due to separation from calves, but was only a result of measurable factors.[4]

"Bovine respiratory disease (BRD), the proper term for "shipping fever," is widely considered the most economically significant disease of fed cattle," noted a 2005 study.[5] The *Angus Beef Bulletin* wrote

---

[2] Saltzman interview with author, Dryden Flight Research Center, Edwards, CA, 5 September 2003.

[3] Larry Cagan, Stanford Research Institute, Menlo Park, CA, to Bud Hartman, NASA Headquarters, Washington, D.C., 24 December 1980, 3, in the private collection of Edwin J. Saltzman.

[4] J. A. Hoffman, D. R. Sandlin, *A Preliminary Investigation of the Drag and Ventilation Characteristics of Livestock Haulers* (Edwards, CA: NASA CR 170408, 1983), 1.

[5] "Bovine Respiratory Disease: A New Look at Causes and Signs of Disease," http://www.mycattle.com/health/updates/brd-nlac.cfm, n.a. (accessed 10 October 2005). "Environmental, nutritional, and management stressors are not primary causes of BRD. Rather, stress causes a rise in blood levels of glucocorticoids, substances which suppress the immune system. As with BVDV, this situation allows pathogens to more easily establish an infection. Environmental stressors include heat, cold (particularly when wind, rain, or mud are involved), dust, and toxic fumes. Nutritional stressors include ration changes, irregular feeding schedules, inadequate access to clean water, and perhaps a need for micronutrient supplementation. Management stressors are numerous. Weaning, transport, co-mingling, crowding, and processing are some of the most common. Damage to the respiratory tract caused by respiratory viruses disarms the animal's innate defense mechanisms. As a result, bacteria that are present in the respiratory tract are allowed to grow and establish an infection. (A cut or scrape, for example, is more likely than intact skin to become infected.) Then, if the animal is stressed, its ability to overcome the infection is weakened, and the BRD process begins." See http://www.mycattle.com/health/updates/brd-recognition.cfm (accessed 10 October 2005). "Recognition and Treatment of Bovine Respiratory Disease Complex," John F. Currin and W. Dee Whittier, Extension Specialists and Professors, Virginia-Maryland Regional College of Veterinary Medicine, Virginia Tech, Publication Number 400-008, posted August 2000.

A conventional tractor-trailor pulling a typical livestock hauler. The tractor makes little to no concession to aerodynamics.
NASA IMG 5150

that in 1991 BRD alone cost cattlemen $624 million.[6] A disease with multiple contributing factors, its recognition and treatment is important to the economic well being of cattle producers. Factors in the disease's development include shipping, weaning, mixing cattle from multiple sources, weather, nutrition, and several respiratory viruses. The cause is linked to environmental stress that seems to trigger onset. Though the cattle carry the mycoplasma (the source of the disease) throughout their lives, the bacteria remain largely innocuous until circumstances change for the host. The bacteria "awake" when the animal is under stress, such as during shipping or when placidity is heavily taxed by some other affliction, and its immune system is weakened. The disease often shows up in nursing calves and in animals gathered in feedlots where they wait in huge numbers for shipping to slaughterhouses. And, of course, it manifests in cattle en route to slaughterhouses.

All these circumstances can result in a calf developing pneumonia, and calves are usually the first, though not the only ones, to succumb.[7] The clinical signs are usually seen 7-21 days after the calves are bought, but can occur anywhere from 2-30 days after purchase. The most common and earliest recognizable clinical sign of pneumonia in cattle is depression.[8]

With the onset of the disease the cows become depressed, their heads droop, their ears hang, and their backs sway. Their breathing becomes more audible and labored and they go off feed. Those cattle who suffer from the disease—and they can die quickly, as one of the names for the disease suggests—are never again as healthy as before, even if they survive. Studies show that cattle that contract BRD and survive don't grow as much as their healthy counterparts over time; they add less weight in a given span than will normal cows, and as a result, they bring less at auction, as much as $20 less per hundred-weight.

The upshot for Saltzman and fellow researchers was another contract, this time from the Department of Agriculture, to examine the question of trailer ventilation. The Dryden Flight Research Center (for by

---

[6] *Angus Beef Bulletin*, September 2001, 1.

[7] Calves usually develop a bacterial pneumonia most often caused by Pasteurella Haemolitica. Pasteurella Multicida and Haemophilus Somnus are also known to cause pneumonia.

[8] Cattle can become visibly depressed. See http://www.mycattle.com/health/updates/brd-recognition.cfm (accessed 10 October 2005).

Depressed Cows

This illustration shows an energy efficient livestock hauler with improved ventilation and temperature control. NASA Illustration

March of 1976 the center's name had changed again), in turn, let contracts with researchers at two schools, the University of Kansas and California Polytechnic State University, San Luis Obispo.[9]

Looking primarily at the shape of the trailer, at openings into the trailer, and at the fairings on the tractor, researchers from both schools reached conclusions, some of which proved counterintuitive. "The intent with each school was to use the general shape of the FRC's low-drag tractor as means of controlling the nature of the air flow into and out of the livestock compartment."[10] The schools also studied existing livestock haulers for comparison. After establishing drag numbers for the model, they began modifying the conventional tractor, and more significantly, the trailer, to improve the aerodynamics of the latter while staying close to the FRC low-drag truck design.[11]

To begin with, separation airflow meant high drag, for the air around the trailer was chaotic. Air that adheres to the trailer's surface produces lower drag and can be directed and channeled. This air can be forced into the trailer in a planned fashion, benefiting the cattle.[12] Their research showed that they could increase the efficiency of the tractor-trailer aerodynamics while simultaneously improving the livestock's chances of surviving the trip. Doing so required new ways to ventilate the trailer, the best being a simple "ram air" induction coupled with modifications to the trailer sides and ends to direct the flow of air most beneficially. Livestock haulers are covered with holes,

---

[9] See Hoffman and Sandlin, *A Preliminary Investigation of the Drag and Ventilation Characteristics of Livestock Haulers*, and Vincent U. Muirhead, *An Investigation of the Internal and External Aerodynamics of Cattle Trucks*.

[10] Saltzman to Gelzer, notes on manuscript draft.

[11] NASA Technology Note: Energy Efficient Livestock Hauler with Improved Ventilation Temperature Control, NASA Ames Research Center (n.d.).

[12] Hoffman and Sandlin, *A Preliminary Investigation of the Drag and Ventilation Characteristics of Livestock Haulers*, 4, and Vincent U. Muirhead, *An Investigation of the Internal and External Aerodynamics of Cattle Trucks* (Edwards, CA: NASA CR 170400, 1983).

of course, on the assumption that this configuration provides at least adequate ventilation for the cattle either while the vehicle is in motion or while stationary. But the research showed not only the chaotic external airflow around the trailer when in motion, but more significantly, poor, often counterproductive airflow in the trailer, airflow that circulated rather than entered and exited.

The design that resulted from the study was for more than just a new tractor: it included a new livestock hauler as well, for the two needed to work together.[13] The fairing over the cab had ducts that forced air into the trailer, and six large NACA ducts located near the front of the trailer (which was boxed in) channeled air into the trailer as well. Large orifices, looking like portholes, spread out across three-quarters of the trailer's length, all the way to the rear, the roof of the trailer entirely sealed. The result looked like no livestock hauler in existence—cab included. In 1982 Saltzman received a U.S. patent for the livestock trailer, the design of which he played the pivotal role in developing.[14] Furthermore, several reports from these research projects outlined the results, and there the matter lay.

---

[13] This may well be the first time that the tractor and trailer were conceived as integral objects strictly for aerodynamic purposes. Prior to this, the work of Dryden engineers had been focused exclusively on the tractor; truck manufactures were no different. Earlier efforts to meld tractor and trailer into seamless units were made for aesthetic rather than genuine aerodynamic purposes, since the trucks did not move fast enough to benefit from the modified shapes.

[14] "Low-drag ground vehicle particularly suited for use in safely transporting livestock." United States Patent 4,343,506 to Edwin J. Saltzman, 10 August 1982.

Chapter Eight
Laws Change; Physics Doesn't

An unexpected factor in the evolution of aerodynamic trucks came with a change in the rules governing over-the-road trucks and trailers.[1] The federal government first regulated the weight and width of interstate commercial vehicles traveling on federal roads in 1956, but at the time focused on the weight of the vehicles and paid scant attention to length and height.[2] Weight of the tractors and trailers was a concern because of stress produced on bridges and road surfaces over which they traveled. The regulations were part of the Federal-Aid Highway Act of 1956 (formally known as the National System of Interstate and Defense Highways Act), which mandated construction of a national highway network running throughout the United States. Ostensibly a project motivated by military needs (to move equipment and personnel quickly about the country, for Eisenhower had experienced both the haphazard American road system and the German autobahn system and recognized the value of the latter), few ever denied the concomitant economic impact such a network would have. For the first time there would be a pre-planned network of limited-access roads linking cities and states across the nation. And as the federal government was paying for the lion's share of it (the payment ratio was 90/10 for the interstate system, federal government to state government, and 80/20 for intrastate roads), Congress saw to it that travel on these roads would comply with federally mandated standards. The 1956 law did set an absolute length for truck-and-trailer combinations at 55 feet. This meant that to maximize the trailer's cargo capacity—its length—the tractor needed to be as short as possible to keep the total within the maximum allowable length.[3]

The criteria set into law in 1956 remained unchanged until 1975, and then the changes were almost entirely related to gross vehicle weight. This was, not surprisingly, in direct response to the soaring cost of fuel, for the weight increase was an attempt to reduce the per-load cost to truckers. Little changed after that until the Surface Transportation Assistance Act (STAA) of 1982, which required that states permit trucks with trailers as long as 48 feet on both interstate and intrastate highways.[4] The lesser highway system is commonly referred to as the National Network, a series of substantial highways that are, nonetheless, second tier. A seemingly minor segment of the new law barred states from "limiting the overall length of a tractor and 48-foot semi-trailer in combination," recalling that under the previous rule both tractor and trailer were part of the total allowable length.[5]

This small portion of the new law had extraordinary consequences. Until 1982 the original qualifier—both the trailer and the combined tractor and trailer length limitations—made a cab-over-engine tractor the choice for long-haul trucks virtually by default. In 1982 this

---

[1] See: *Comprehensive Truck Size and Weight Study* (Washington, D.C.: US DOT FHA), in four volumes, August 2000. http://www.fhwa.dot.gov/policy/otps/truck/finalreport.htm (accessed 3 June 2009).

[2] Ibid. "A maximum gross weight limit of 73,280 pounds was established along with maximum weights of 18,000 pounds on single axles and 32,000 pounds on tandem axles. Maximum vehicle width was set at 96 inches, but length and height limits were left to State regulation. States having greater weight or width limits in place on 1 July 1956, when Federal limits went into effect, were allowed to retain those limits under a grandfather clause."

[3] *Comprehensive Truck Size and Weight Study*, chapter 2, 5.

[4] Current federal law (sec. 411) requires that states allow a minimum of 48 feet for single trailers, and a maximum of 53 feet (with a few exceptions). This law supersedes state laws governing trailer lengths, although states can make individual exceptions for certain loads. "No state is allowed to set tractor length limits." http://www.geocities.com/thetropics/1608/page11.htm (accessed 9 September 2009).

[5] Ibid., 6. The width of the trailer was increased from 96 to 102 inches, providing the roadway was 12 feet wide. (As of 2005, federal law denies states the right to limit trailer length to less than 48 feet. At this time there is no mandated uniform length for trailers; though the common length is 53 feet, some states, such as Wyoming, allow trailers as long as 60 feet.) This was why the NASA-FRC team applied their modifications to a cab-over rather than a conventional model.

At left, a cab-over-engine tractor, a style increasingly rare on American highways for aerodynamic reasons but one that retains advantages in certain environments. The image at right is of a conventional tractor.
NASA
ED10-0030-8, IMG-5117

truck style constituted over 60 percent of the market for the Peterbilt Corporation, and the numbers were similar for other manufacturers. (The rest of the company's market did not fall under these rules, obviating the cab style.) But since the new rule applied only to the trailer, it effectively removed any length limitation to the overall combination (to a point). The tractor now no longer needed to be the cab-over type. In turn, this allowed engineers to apply aerodynamic improvements to an already superior truck shape, the "conventional" cab.

The difference between the two kinds of truck cabs is quite material when discussing aerodynamic drag. Cab-over-engines, or simply cab-overs, place the cab directly above the engine, minimizing the length of the tractor. This results in a cube-like tractor efficient in its use of linear space but with little that can be done about the large, flat frontal area. Conventional trucks, on the other hand, place the engine ahead of the cab and are longer as a result. Locating the engine ahead of the cab enables designers to create a more pointed shape for the truck, something that directs the air more efficiently than the billboard-like front of a cab-over design.[6]

One of the most dramatic effects of this change in the 1982 law came in the rising popularity of conventional cab tractors, once tractor designs were no longer restricted by laws governing length relative to the trailers being pulled. Unconfined by overall lengths imposed by the federal government, truck manufacturers turned with a vengeance toward conventional trucks and their inherent aerodynamic potential, and the cab-over model gradually fell out of favor. For the Peterbilt Corporation the cab-over design represented 65 percent of sales in 1980, but only 1 percent by 2004.[7]

Perhaps more than anything else, this change in the law redirected the market for long-haul trucks toward conventional models, and created even more emphasis on the need for design of tractors with more aerodynamically efficient shapes. What began with the Kenworth T600 blossomed into a wave of new truck designs from all major American truck manufacturers seeking to keep abreast of a changing market. Moreover, the modifications tried by engineers at Dryden were adopted by truck manufacturers, as were the lessons derived from the road tests conducted at the center, for what the NASA engineers demonstrated with cab-overs applied to conventional models as well as to smaller, short-haul trucks. The cargo boxes of most delivery trucks today have rounded front corners and edges, a direct application of the research conducted at Dryden on the Shoebox. A glance at any modern long-haul tractor-trailer unit will show Dryden's influence there as well, in the sweeping fairing from the cab up to the trailer's roofline to the narrowing (if not elimination) of the tractor-to-trailer gap, in the appearance of boattail-like structures on some of trailers, and in the effort to seal the gap between trailer and ground.

---

[6] Relocating the engine forward and lengthening the overall tractor wheelbase also had the beneficial side effect of lessening the jolting ride of the COE.

[7] Derek Smith, PACCAR, Inc., electronic mail correspondence with author, 16 September 2003.

Chapter Nine
The Drag Bucket

Configuration*	Corners Front	Corners Rear	Underbody
A	Square	Square	Exposed
B	Vertical rounded, horizontal square	Vertical rounded, horizontal square	Exposed
C	Rounded	Rounded	Exposed
D	Rounded	Rounded	Full-length seal
E	Rounded	Rounded	Three-fourths-length seal
F	Rounded	Square	Three-fourths-length seal
G	Rounded	Full boattail	Full-length seal
H	Rounded	Truncated boattail	Full-length seal

*Configuration F approximates configuration I of reference 21 except that configuration I has a full-length underbody seal (fairing). Configuration G is configuration II and configuration H is configuration III of reference 21. Drag coefficients for configurations G and H are averaged for V = 50 mph and 60 mph. Drag coefficients for configurations A to F were obtained at V = 60 mph.

Configuration	C_D, NASA Dryden–designated	$\frac{\Delta C_D}{C_{D_A}}$, percent	C_D, conventional
A	1.13	0	0.89
B	0.68	40	0.54
C	0.520	54	0.410
D	0.440	61	0.347
E	0.443	61	0.350
F	0.463	59	0.365
G	0.302	73 (.733)	0.238
H	0.307	73 (.728)	0.242

These two tables enumerate baseline and all modifications to the Shoebox van, as well as the results.

In 1997 the U.S. Department of Energy sponsored a transportation-related workshop in Phoenix, Arizona, to which Dryden representatives were invited because of their earlier work. Those from the center who attended, including Saltzman, came away somewhat perplexed at the industry's expectations, for the stated goals of the trucking industry seemed in conflict with its actions, if not also with existing research. "We want," said a representative at the meeting, "an efficiency somewhere between a DC-3 and a Pontiac Firebird." Specifically, the agreed-upon goal among trucking industry representatives was a drag coefficient of .25 for the tractor-trailer combination.[1]

Hearing this, the group of NASA aerodynamicists that attended the meeting was somewhat surprised. The best any of their tests had achieved was a Cd. of .242, and though this was slightly below the industry goal of .25, that figure came only with the Shoebox completely faired in the front, the underbody and wheel wells sealed, and the afterbody sporting a truncated boattail. The best they had achieved with the modified tractor-trailer unit was a Cd. of .59, although that was with no modifications to the trailer's aft end. And yet the long-haul trucking industry now expected to achieve virtually the same low Cd. Saltzman's team had with the Shoebox.[2]

Following the meeting the NASA team noted that in order to reach this goal the trucking industry would have to modify not only the cab area, which it had been doing for nearly two decades, but would have to address the trailers as well. Fairings (i.e., boattail-like features) would need to appear on the back end of the trailers to smooth the airflow while on the road, and the bottom of the trailer would need considerable attention. But it was precisely what the trailers needed most of all that the manufacturers and shipping companies would likely avoid: addition of a boattail to raise the pressure at the base. Without all this, said Saltzman, the goal of Cd .25 would be unreachable.

Saltzman then turned his attention once again to the question of forebody and base pressures (he had been focused on other things in the years since the early

---

[1] Writing in the early 1960s, aerodynamicist John Allen pointed to a steady reduction in automobile drag reduction over the decades, but he admitted uncertainty over whether this was a result of fashion or of function. John E. Allen, *Aerodynamics: The Science of Air in Motion* (New York: McGraw-Hill, 1982).

[2] One recent study argues: "Excluding pneumatic blowing, [the] theoretical limit for the coefficient of aerodynamic resistance for combination trucks" is in the range of .13 to .19, while admitting that the best current tractor trailer combinations of the period are in the .6 to .7 range. The study's authors concede that realistic drag coefficients in the best of circumstances might be .25. *Winning the Oil End Game*, Technology Annex chapter 6, "Class 8 Heavy Trucks," 6: www.oilendgame.org (accessed 18 June 2009).

The Drag Bucket

truck fairing tests), and to earlier research conducted at the center. One of the reports he help author laid out the trucking industry's problem with respect to the stated goal of a Cd. of .25: "Two versions of the test van ground research vehicle (configuration A and F) demonstrate that as forebody drag is reduced, the base drag is increased."[3] Here on paper, graphed with empirical data, was the evidence illustrating his group's puzzlement following the 1997 meeting in Phoenix. Aerodynamic improvements to the front had to be matched by similar improvements to the back or there would be only limited gains. Simply put: the industry would never see its goal without dramatic change to the trailer's aft end.

While thinking again about the issue of fore and base drag Saltzman had an epiphany. What he realized, ironically, was that efforts to reduce forebody aerodynamic drag had certainly yielded dividends, but those dividends were inevitably finite. He concluded that beyond a certain point the aerodynamic refinements to the truck's forebody would generate more drag, not less. Looking at the graphs of earlier Shoebox research more carefully, Saltzman noticed that this "rule" of limited return resulted in what he dubbed a drag "bucket" for blunt-shaped vehicles.[4] The extraordinary revelation surfaced formally and publicly in a paper he, K. Charles Wang, and Kenneth W. Iliff

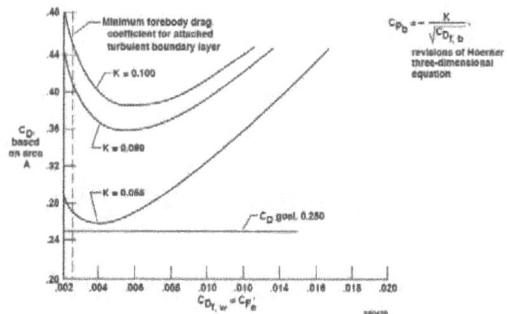

The first published graph of Saltzman's "drag bucket." Here the term refers to finite gains achievable through aerodynamic refinements to fore and aft areas of a vehicle, after which drag inexorably rises.

presented in 1999. In it the authors noted of the family of aircraft with blunt forebodies and flat afterbodies: "The salient feature of these curves is that each has an optimum region of lowest overall minimum drag coefficient."[5] Regardless of what changes were made to the fore and aft of a vehicle to improve its aerodynamics, there existed a point of maximum return—of minimum drag—after which drag actually rose again. The three researchers had gone back and looked at the drag data from the family of lifting bodies flown at the FRC,

---

[3] The paper goes on to suggest that the necessary modifications to the trailer's aft end might well lead to the reappearance of cab-over-engine trucks in order to accommodate the increased length of the overall vehicle. Edwin J. Saltzman and Robert M. Meyer, *A Reassessment of Heavy-Duty Truck Aerodynamic Design Features and Priorities* (Edwards, CA: NASA/TP 1999 206574), 25. Appropriately, Saltzman and Meyer had been the authors of the first formal request to the FRC director for approval of research and development "to study the efficiency of ground vehicles," made in November 1972.

[4] The term "drag bucket" is not unique to Saltzman and this project. In one classic text for aviators, for example, it is used to indicate the point of greatest laminar flow, and therefore lowest induced drag, of a particular airfoil. H. H. Hurt, Jr. *Aerodynamics for Naval Aviators* (reprint: Aviation Supplies & Academics, 1965), 54-55. An alternate term is "minima," suggesting a low point on a graph.

[5] Edwin J. Saltzman, K. Charles Wang, and Kenneth W. Illif, *Flight-Determined Subsonic Lift and Drag Characteristics of Seven Lifting-Body and Wing-Truncated Reentry Vehicle Configurations with Truncated Bodies* (AIAA 99-0383, 1999), 27.

along with the X-15 and the space shuttle, the "generic blunt-based class of vehicles," and found precisely what Saltzman had earlier realized existed.[6] There was a point after which improvements to forebody drag were accompanied by large increments of base drag. Therefore, a minimum drag value would exist for this class of vehicle. And while there was visual evidence of this before Saltzman's epiphany, he was the first to realize that data existed to explain this phenomenon, and to offer evidence for why this was happening.[7]

This raised the specter of never seeing a drag coefficient of .25 for the trucking industry unless a boattail were used, drastic changes were applied to wheel and undercarriage of the trailer, or other exotic treatments that were unlikely to be practical in application.[8]

---

[6] Ibid., 26. What photos of fluid in motion had long showed, remarked Dryden aeronautical engineer Albion Bowers, Saltzman provided the evidence to explain: after cleaning up the front and back of an object there wasn't enough energy in the flow to keep that flow attached to the object. Albion Bowers interview with the author, NASA Dryden Flight Research Center, Edwards, CA, 4 December 2009.

[7] See, for example, Milton Van Dyke, *An Album of Fluid Motion* (Stanford, CA: The Parabolic Press, 1982; reprint 2007), 34-35.

[8] Results from the FRC tests of aftermarket add-on aerodynamic devices showed that the best device, Rudkin-Wiley's Airshield, produced a Cd. .89. This was without any additional work, such as addition of trailer side skirts or a boattail or narrowing the tractor-to-trailer gap, but it is nevertheless indicative of the numbers the trucking industry had to contend with. The best Cd. that FRC researchers achieved with the Shoebox—but without the boattail on that vehicle—was .52. Saltzman, *A Summary of NASA Dryden's Truck Aerodynamic Research*, 4.

In 2009 three authors published a paper with the Society of Automotive Engineers covering research conducted with truck models in one of NASA Langley's former wind tunnels. Starting with a baseline, unmodified model, the team applied six aftermarket aerodynamic applications meant to improve truck efficiency. Of those they tried, the only ones that were entirely different from those tried by researchers at the FRC in the mid-70s were "skirts," rigid slabs of material added to seal the area between the cab's rear wheels and the trailer's wheels. The seventh configuration tested in the wind tunnel had a full skirt (covering the entire trailer's length, including the cab's wheels), the entire gap between tractor and trailer sealed, and a boattail attached to the trailer's aft end. Not surprisingly, this was the most efficient of the configurations. Most significant, however, is that this configuration yielded a Cd. of roughly .37 at 55 mph and almost the same at 65 mph, and this was with a truck model sporting all the available modifications, including a boattail. Drew Landman, Richard Wood, Whitney Seay and John Bledsoe, *Understanding Practical Limits to Heavy Truck Drag Reduction*, #2009-01-2890, Society of Automotive Engineers, 2009.

Chapter Ten
Results

Where has all this research led?

Historians of technology have argued for some time that technology does not exist or develop in a vacuum; it is advanced, inhibited, or curtailed by cultural influences.[1] The path from bulky and brawny to sinuous, aerodynamic tractors is such a case. New technology, for example, is rarely met with open arms since it represents a threat of some sort, and aerodynamic trucks were no exception.[2] Resistance surfaced early.

When NASA drivers took the FRC's highly faired tractor-trailer out on its 312-mile loop, they paused at truck stops for lunch and in order to slyly show off the truck. One of the NASA drivers, himself a former professional driver, mingled with other drivers, for, in Saltzman's words, "he knew the language." Some would come out to see the vehicle, and listen to the benefits the aerodynamic refinements generated in fuel savings, which were very real. "You won't catch

---

[1] Consider, for example, air travel. Heavier-than-air flight is barely a century old as of this writing, yet it has advanced dramatically in that time. Piston engines moved passengers at less than 300 mph prior to World War II; in the years following the war that speed jumped to 400 mph, still with piston engines. The introduction of the turbojet meant that passengers could fly at 440 mph in a DeHavilland Comet. Meanwhile, piston engine military aircraft that, before the war, could nibble at the transonic realm were replaced after the war with aircraft that could exceed the speed of sound. By 1953, the year in which Douglas introduced the DC-7, considered by many to be the pinnacle of piston-engine propeller airliners (cruise speed of 405 mph), Scott Crossfield exceeded Mach 2. Before that decade was out some military aircraft were routinely flying at Mach 2. Boeing's 707, which entered service in the same decade, cruised at just over 600 mph. Everything pointed to ever-increasing speed in transportation, and indeed, aircraft with air-breathing engines began exceeding Mach 3 in the early 1960s, and passengers were regularly crossing the Atlantic in sumptuous luxury at Mach 2 by the late 1970s. Yet at the beginning of the twenty-first century commercial airliners operate in the same realm as they had 50 years earlier: subsonic. More interesting is that the culture in this respect has actually slowed down; Mach 2 commercial service is no longer available. This seeming oddity is not a function of technology, but of social forces: cultural values rather than any logical progression determined by the course of technological development and its application. The literal cost of constructing and operating commercial supersonic aircraft, the impact of sonic booms on the pedestrian public, and other factors combined to make Mach 2 flight the preserve of the military in the early part of the present century.

The Douglas DC-4 (1942) cruised at 265 mph; the DC-7 (1953) at 405 mph; and the DeHavilland Comet 1 (1952) at 440 mph. The Aerospatiale Concorde, which entered commercial service in 1976 and operated until 2003, cruised at roughly Mach 2. Various aircraft manufacturers have taken up the cause of supersonic commercial flight recently, if only in terms of a business jet, and NASA has been a contributor to new technology for civilian supersonic aircraft, but nothing has materialized as of this writing. Regarding factors concerning the failed U.S. supersonic transport intended to compete with the Concorde, see: Erik M. Conway, *High-Speed Dreams: NASA and the Technopolitics of Supersonic Transportation, 1945–1999* (Baltimore: Johns Hopkins University Press, 2005). On the larger subject of the social construction of technology, see Wiebe E. Bijker, Thomas P. Hughes, and Trevor J. Pinch, eds., *The Social Construction of Technological Systems: New Directions in the Sociology and History of Technology* (Cambridge: MIT Press, 1987); Donald MacKenzie, *The Social Shaping of Technology* (Open University Press, 1999); and Merritt Roe Smith and Leo Marx, eds. *Does Technology Drive History? The Dilemma of Technological Determinism* (Cambridge: MIT Press, 1998) as merely three titles. Many more exist for the reader wishing to delve further into the subject.

[2] Historians working under the rubric of the social construction of technology operate from an essential premise: the "'interpretive flexibility' of an artifact." The idea is that different groups vest the same artifact or technology with different meanings. "For young men riding the bicycle for sporting purposes, the high-wheeler meant the 'macho machine' as opposed to the meaning given to it by women and elderly men who wanted to use the bike for transportation. For this latter group . . . the high-wheeler was the 'unsafe machine' because of the habit of throwing people over the handle bars—known as 'doing a header.'" Ronald Kline and Trevor Pinch, "The Social Construction of the Automobile in the Rural United States," in Merritt Roe Smith and Gregory Clancey, *Major Problems in the History of American Technology* (Boston: Houghton Mifflin Company, 1998), 337. The tractor that, to the owner-operator, was an extension of said owner-operator was to a trucking firm merely a tool of no particular social significance. Therefore, the subject of a truck's manliness was of greater or lesser concern, depending on which group one belonged to.

me driving anything like that," they'd say, "it's a sissy truck."³

The persistence of old technology alongside the new is another characteristic of technological development; new and improved does not sweep away the old, however compelling the logic.⁴ The well-faired shape of today's long-haul tractor, though increasingly common, is matched by a stubborn persistence of old shapes to survive, right down to the external location of air filters and flat front of the Peterbilt model 379. When asked about this, a company representative replied: "Does the operator want a truck that looks like a truck or one that looks more like the current cars and SUVs? The 387 is inherently more aerodynamic than the 379, but there are a significant number of buyers who want classic styling, with a big engine, and are willing to accept lower fuel mileage."⁵ This attitude made truck manufacturers reluctant to radically change their products—at least until motor freight carriers did the math.

But change they did, with Kenworth Truck Company leading the effort among manufacturers and aftermarket firms offering a variety of fairings to smooth the truck and trailer's passage through the air. A 2005 Kenworth Company white paper bluntly noted the unyielding problem facing anyone operating a long-haul truck: "Approximately half the energy used by a truck traveling at 55 mph is simply to move the air around that truck. At 65 mph, about two thirds of the energy is used to cut through the air."⁶

In fact, nearly a decade elapsed between NASA's early experiments and the first truck manufactured to new aerodynamic designs. In that interim those who wanted aerodynamic improvements had no choice but to turn to aftermarket suppliers—firms such as NoseCone FitzGerald or Airshield that had been sell-

---

³ Saltzman interview with author, 5 September 2003. This is echoed in comments made by truckers in 2005 when surveyed about their preference for "traditional" or "aerodynamic" truck styling. Asked for his preference, Melvin Mills responded: "Regular. The newer ones look like something out of Star Wars." "Do You Prefer Traditional or Aerodynamic Trucks?" *Overdrive*, June 2005. http://www.etrucker.com/apps/news/article.asp?id=47579 (accessed 3 June 2009). There is no doubt that vehicles (and technologies in general, as well as professions) are gendered, and not just in American culture. Ships and, later, airplanes have traditionally been denoted as female. Trucks have often been gendered male, although there is no explicit tradition for this. Americans have even gendered certain types of automobiles: in 2001 a university student was pulled over by an Alabama state trooper because he was driving a Volkswagen New Beetle, which the officer deemed a "girl's car." The officer wanted to know why the young man was driving the "wrong" car. Dr. Stephanie Smith, Auburn University History Department, interview with the author, April 2002. Some of this author's students agreed that the New Beetle was a girl's car because it has a small vase that is part of the dashboard. More examples of vehicular gendering can be found in newspaper articles discussing "lesbian" and "gay" cars, which include the Subaru Outback, the Mazda Miata, and the new Mini Cooper (the assigned gender of which seems to depend on whether or not the car is a convertible).

⁴ Examples of this abound, including propeller aircraft that continue to fly alongside jets seventy years after jets were introduced. Taking one case in more detail, the whaling ship *Charles W. Morgan* first went to sea in 1841 and continued whaling under sail until its retirement in 1921. Yet 22 years before the *Morgan's* launch, the *Savannah* became the first ship to cross the Atlantic with steam power, in 1819 (it could not carry all the necessary coal to make the trip entirely under steam and was under sail much of the way, but the point was made), and vessels were plying the Atlantic entirely under steam while the *Morgan* was still a young vessel. By the end of the nineteenth century large ocean-going steamships relied on complex triple and even quadruple expansion steam engines to capture as much energy from the steam as possible. In the midst of this increasingly sophisticated steam engine development, Charles A. Parsons launched his *Turbinia* in 1894, the first vessel to have steam turbines for propulsion (axial flow). He made the biggest splash with his invention at Queen Victoria's Diamond Jubilee, in 1897, when he truculently drove his small craft around the numerous British Royal Navy vessels with impunity, knowing full well that nothing afloat could catch him as the *Turbinia* could do 34.5 knots. Though deeply embarrassed in front of the queen and foreign dignitaries attending the event at Spithead, the Lords of the Admiralty swallowed hard in the face of the obvious: they ordered two Parsons-built turbine-powered destroyers that were launched in 1899. The first major battleship, the HMS *Dreadnought*, was launched in 1906 and was propelled by steam turbines. The *Dreadnought*, whose very name defined an entire category of naval ships and whose technology signaled a fundamental change in propulsion, was retired in 1922, just a year after the sailing ship *Charles W. Morgan*. See: Phil Russell. *Navies in Transition: The Development of the Worlds Navies, the Technology, and the People Who Made It Happen.* "Sir Charles A. Parsons, 1854-1931." http://www.btinternet.com/~philipr/Parsons.htm (accessed 29 September 2009), and http://www.mysticseaport.org/index.cfm?fuseaction=home.viewPage&page_id=B3E6C64-B3CA-45AE-A83D72C303A9C6BF (accessed 9 September 2009).

⁵ Derek Smith, PACCAR, Inc., electronic mail correspondence with author, 16 September 2003.

⁶ "White Paper on Fuel Economy," Kenworth Truck Company, October 2001, 3. It's worth remembering that this figure comes from 2005, by which time long-haul truck had seen the benefits of more than 20 years of aerodynamic refinements.

ing aftermarket aerodynamic devices since 1968 but with little initial success. "In the beginning very few truckers were interested," recalled the president of Airshield, N. C. Wiley, in 1974. "Diesel fuel at the time was 22 cents per gallon and just about the last thing anyone worried about. The cost has now doubled and sales of aerodynamic devices have tripled."[7]

Despite the fact that NASA's work was on a cab-over and not a conventional model, ongoing efforts within the trucking and the truck aftermarket industries continue to echo the work originally done at the Flight Research Center in the 1970s, or work the center coordinated with universities.[8] In 2006, for example, the AB Volvo Truck Company announced three improvements to one of its tractors, changes designed to boost efficiency. One reflected the original work done by Saltzman's team: sealing much of the underside of the tractor to smooth airflow beneath the vehicle.[9] Or consider a study conducted by the Mack Truck Company, now a subsidiary of AB Volvo. Following two years of research between 2004 and 2006 the organization found that "significant fuel savings can be achieved by enclosing the gap between the tractor and trailer, and equipping the trailer with what are referred to as 'side skirts' and a 'boat tail' to improve its aerodynamic profile while on the road."[10] And the firm Parker-Hannifin, a diversified manufacturer with interests in transportation, underscored the validity of the FRC work when it joined the Get Nitrogen Institute in calling for the use of nitrogen to inflate tires, the better to maintain proper inflation on the road, thereby improving mileage.[11]

And even with the design changes made by truck manufacturers, the aftermarket field remains active. Several firms continue to offer solutions to the problem of aerodynamics, something that suggests a continued reluctance by truck manufacturers to lead the march to efficiency, perhaps mindful that radical changes can evoke quick rejection by customers. One example involves Aeroserve Technology Ltd., of Ontario, Canada, which sells vortex generators that attach to the side of the trailer just ahead of the back end, running from the top of the trailer to its bottom. Called AirTabs, the products are designed to generate vortices that energize the air as it is forced to break away from the trailer; doing so mitigates the abruptness of the pressure differential and helps reduce the developing low-pressure zone at the aft end, reducing drag.[12] And NoseCone FitzGerald continues to offer a variety of add-ons for both long-haul and daily-delivery trucks.

For all the modifications available from truck manufacturers and aftermarket suppliers, however, the desire

---

[7] "Demonstration Aerodynamic Drag Reduction for the purpose of Reducing Fuel Consumption of Trucks," speech given by N. C. Wiley, President, Airshield Division, Rudkin-Wiley Corporation (Stanford, CT, 5 December 1974), 13. In 2009, at the time of this monograph's writing, diesel fuel prices had already neared $5 per gallon in California.

[8] A 2009 study conducted with a truck model and aftermarket fairings in the former Langley wind tunnels, and reported in an SAE paper, acknowledges the University of Maryland's Trailmobile study, but none of the work done by NASA between 1973 and 2000. The oldest source the 2009 study's researchers cite is another SAE paper, from 1985. Yet, with one exception, every model configuration that team tested in the wind tunnel had either been tested by NASA or its partners in that earlier span of 27 years. See Landman, et al., *Understanding Practical Limits to Heavy Truck Drag Reduction*, #2009-01-2890. For comparison, see Muirhead, *An Investigation of Drag Reduction for Tractor Trailer Vehicles* (NASA CR-144877), October 1978. With respect to the matter of cab-overs vs. conventional tractors, Saltzman noted (ironically) in 1999 that "designers may want to again use a type of cab-over-engine tractor so that the shorter wheel base can afford extra trailer length devoted to base drag reduction devices." Saltzman and Meyer, *A Reassessment of Heavy-Duty Truck Aerodynamics* (Edwards, CA: NASA TP-1999-206574), 26. Perhaps no designer has taken on the aerodynamic cab-over challenge with more imagination than Swiss-German designer Luigi Colani. See http://www.autoblog.com/2007/01/02/radical-semis-by-luigi-colani/(accessed 11 July 2011).

[9] "Volvo Displays Aerodynamic Devices For Improved Fuel Economy," 15 November 2006. http://trailer-bodybuilders.com/news/volvo_fuel_economy/ (accessed 3 June 2009).

[10] "Mack Research Shows Potential 8% Fuel Economy Improvement," 15 November 2006. http://www.truckinginfo.com/news/newsdetail.asp?news_id=57555&news_category_id=20 (accessed 11 January 2007).

[11] "Parker Hannifin Joins Get Nitrogen Institute," 15 November 2006. http://www.truckinginfo.com/news/newsdetail.asp?news_id=57542&news_category_id=20 (accessed 11 January 2007).

[12] "Expediter Leo Bricker says that he believes the AirTabs he has installed on his Kenworth W9 chassis D straight truck have resulted in enhanced stability and reduced buffeting, especially when it is being passed by another vehicle. Bricker says that improved rearward visibility in rain and snow is another plus. 'On a truck without AirTabs, all I can see behind me in the rain is a foggy spray. With the AirTabs, I have a clear field of vision back there.'" http://www.expeditersonline.com/artman/publish/truck-aerodynamics.html (accessed 3 June 2009).

for style continues to be a major factor in the trucks that owner-operators, and even fleet owners, buy, usually at the expense of fuel economy. "Owner-operators want distinctive styling—not necessarily classic styling," admitted Bob Weber, an engineer with International Truck and Engine Corporation. "These are the Pride & Polish guys. They're hardcore image guys." And in 2005 the Mack Company, the same firm that conducted a two-year research program to find ways of improving truck efficiency, launched its "Rawhide" model, with owner-operators as the target market. The campaign slogan: "Legendary performance meets classic style." Among the features defining the truck is a "Texas bumper" (a large, flat, chromed bumper), 6-inch chromed exhaust stacks (larger in diameter than stock exhaust stacks), and air horns mounted on its flat roof. None of these make any concession to aerodynamics. And as a Kenworth study demonstrated, the difference between a truck with classic styling and a highly faired style could be as much as 15 percent in fuel efficiency.[13]

Despite the persistence of "classic" styling, the application of these collective experiences and experiments is evident in the majority of new over-the-road tractors. Today's trailers, on the other hand, look little different for the intervening years. This is, in part, because the optimal shape for superior aerodynamics on the aft end of a trailer—the boattail—must be one that is functional, easy to open and close, and impact-resistant to a point, since the trailers are regularly backed up to loading docks. Unless they can fit all these criteria, boattail fairings are unattractive to trailer manufacturers even though they would produce the greatest improvement in efficiency.[14]

That said, the back end of the trailer garners increasing attention from aftermarket manufacturers. Washington state-based Aero Works, for example, touts its "new approach to 'boat tails.'" Noted Wayne Simons, a Kenworth engineer then working with Aero Works on a joint project in 2009: "Our testing has shown that their concept boat tail can improve fuel economy by several percent" because of a lower aerodynamic drag coefficient.[15] And in a version of "what's old is new again," in 2001 researcher Bob Englar of the Aerospace, Transportation & Advanced Systems Laboratory at the Georgia Institute of Technology, working in conjunction with the U.S. Department of Energy and the American Trucking Association, explored an idea for aerodynamic improvements through active boundary-layer control. The plan, dubbed circulation control, involved blowing air over surfaces of the trailer, particularly the trailer's top edge. According to researchers, this smoothed airflow (reduced its turbulence) and led to as much as a 15 percent drop in the weight of the trailer by generating lift, improving vehicle efficiency.[16] In that same spirit, the splitter plate with which FRC engineers experimented on the center's tractor-trailer combination (and that AeroVironment also tried out on its standard truck experiments) has re-appeared in a slightly modified form as the "cross-flow vortex trap device." Instead of one single plate filling much of the gap between cab and trailer, SOLUS-Solutions and Technologies is seeking a patent for six or seven smaller plates, attached at the trailer's front end, that are meant to trap cross flow by generating small vortices between pairs of plates. The firm also has devised a set of strakes (Vortex Strake Devices) that attach to the aft end of the trailer at angles. The theory is that as air flows over the angled strakes a vortex begins, swirl-

---

[13] *Overdrive*, June 2005. http://www.etrucker.com/apps/news/article.asp?id=47566 (accessed 3 June 2009).

[14] Some European manufacturers have started building trailers with tapering rooflines at the back of the trailer, with the expectation that this will guide airflow into the low-pressure zone. Additionally, at least one British firm has also tapered the front of the trailer's roof to fair it into the cab's air deflector. *Transport Energy Best Practices: The Streamlined Guide to Truck Aerodynamic Styling, Department of Transport* (UK) (Crown, 2004), and *Transport Energy Best Practices: Smoothing the Flow at TNT Express and Somerfield using Truck Aerodynamics* (n.p., n.d.).

[15] "NOW THAT'S A BUMPER," March 2003. http://www.etrucker.com/apps/news/article.asp?id=34858 (accessed 3 June 2009).

[16] This calls to mind the 1936 patent for a car with its own boundary-layer control mechanism. John Toon, "Flying Low-Drag Trucks: Aerodynamic concepts and controls for aircraft will cut fuel use and improve control in trucks," Georgia Tech Research Horizons, 16 February 2001. http://gtresearchnews.gatech.edu/reshor/rh-win01/trucks.html (accessed 3 June 2009).

[17] Richard M. Wood and Steven X. S. Bauer, *Simple and Low-Cost Drag Reduction Devices for Tractor-Trailer Trucks*, SAE 2003-01-3377, 2003.

These two images show the second, stock tractor, ready to pull the trailer ordinarily pulled by the modified tractor (note the three horizontal smears extending from the front vertical edge of the trailer, evidence of the rubber rollers on the gap seal panels). Notable in both images is the splitter plate applied by the fabrication shop to the trailer's front face. The device was meant to help close the gap between tractor and trailer, reducing crosswind effects and the resulting aerodynamic drag.
NASA   *E76-30153, E76-30170*

ing into the base area of the trailer and raising the base pressure.[17]

Another aftermarket firm with its eye on the aft end of the trailer at the time of this writing was Freight Wing, Inc., which offers tractor-to-trailer gap, belly, and rear fairings for existing trailers. The fairings attach to the bottom outer edges of the trailer, helping to close off its large, open underside. Gordon Trucking, Inc. had applied Freight Wing's trailer side skirts to some 1,650 of its trailers, or about a third of its fleet, by mid-2010. The entire fleet will have side skirts when the program is complete. The aerodynamic change yields roughly a 3.2 percent improvement over the standard truck's fuel economy, said Kirk Altrichter, GTI vice president of maintenance.[18] More unusual, however, is Freight Wing's rear fairing; a set of curved

Freight Wing, Inc., offers several aftermarket products for trailers that are gaining popularity, items that reflect research conducted by NASA and universities. Among them is the side skirt that cuts down on cross flow beneath the trailer, as well as turbulence.
Image courtesy Freight Wing, Inc   *IMG-7836*

A closer view of a trailer side skirt. Freight Wing, Inc., which manufactured this side skirt, offers a range of products to reduce aerodynamic drag on tractor-trailer units, as do several other manufacturers.
Image courtesy Freight Wing, Inc   *Wh-077*

---

[18] Kirk Altrichter, telephone interview with the author, 19 July 2010. Altrichter noted that SAE and government studies of side skirt-equipped tractor-trailers identified fuel savings of 7.4 percent on average, but that was only on closed-circuit test tracks, whereas Gordon Trucking International numbers reflected actual over-the-road driving experiences carrying freight.

Results

A late model Gordon Trucking, Inc. tractor-trailer unit featuring a well-faired cab and trailer side skirts.
NASA   ED10-0030-9

plates attach to the rear doors to help round off the trailer, a modification similar to the Shoebox's boattail. The distinction, however, is that Freight Wing's fairing collapses quickly so that the trailer's doors can be opened and pushed flat against its sides during loading at a dock. The benefit claimed when all three devices are used together is a 7 percent reduction in fuel consumption over existing, unmodified trailers. Moreover, the products were designed to fit but not interfere with existing and new truck models outfitted with their own aerodynamic improvements. A patent was issued in 2004 for a similar arrangement, although the collapsible boattail in this instance is much more pronounced, more closely resembling the final Shoe-

[19] U.S. Patent 6799791—Deployable vehicle fairing structure, issued on October 5, 2004, for "a compact rear fairing to reduce the drag incident to relatively high speed movement of box-like bodies, such as trucks, trailers and cargo containers, is provided. The structure of the fairing is substantially rigid and, depending on the use thereof, is formed with two or more outer surfaces shaped in the contour of the upper surfaces of an air foil, the leading surfaces of which are mountable adjacent the rear of the box-like body and the trailing surfaces thereof being joined together to form an apex at its rear." The fairing is readily mounted on and detached from the box-like bodies and interchangeable among trucks, trailers, and containers of the same size. As of this writing no fewer than 41 patents have been issued for various vehicle aerodynamic improvement devices, at least nine since 2001. Quite a few are for collapsible boattails (one resembles a pyramid tipped on its side). Several patents are for devices attached to the trailer that direct air from the sides of the trailers into the low-pressure zone at the trailer's aft end, including "pipes" and "scoops." See for example U.S. patents 3,371,146; 3,929,369; 3,999,797; 4,095,835; 4,102,548; 4,131,309; 4,142,755; 4,316,630; 4,458,936; 4,601,508; 4,702,509; 4,978,162; 5,058,945; 5,108,145; 6,092,861; 6,309,010; 6,409,252; 6,742,616; 6,854,788; and 7,216,923. The last patent in the list involves piping engine exhaust to the trailer's back end with the hopes of raising the pressure in this area, reducing aerodynamic drag. An interesting one from 1978 (4,095,835) is for a "deployable streamlining" feature that resembles an umbrella opened in front of the truck cab so that the tip of the "umbrella" points in the direction the truck is moving. This is just a short list of patents for aerodynamic improvement devices targeting trucks. http://www.google.com/patents?id=thGqAAAAEBAJ&dq=6467833 (accessed 3 June 2009).

box configuration.[19] And as with Kenworth's T600, the first genuine aerodynamic long-haul tractor, fleet operators are playing a substantial role in popularizing trailer side skirts.[20]

For livestock haulers, the investment in the redesign of trailers renders manufacturers disinclined to revamp their systems. One factor in this is that, since deregulation of the livestock hauling industry in 1975, individual farmers/ranchers have become the predominant owners of the "cattle pots," and these owners benefit little in the cost of a redesigned trailer given their infrequent use and purchase costs.[21] It is not entirely unreasonable a reaction given that each cattle rancher may own only a pair of trailers and they remain parked for most of the year. Makers of livestock trailers have so far been indifferent to the suggested reshaping that Saltzman and his colleagues recommended, essentially because their market has demonstrated no interest in the changes. "And this is in spite of the data from Muirhead, Hoffman, Sandlin, and the FRC, all of which showed the significant improvement the proposed new shape and ventilation exhibited when compared to existing livestock haulers."[22] Of course, trailer makers have no reason to redesign their trailers so long as trucks don't incorporate the designs that must go hand in hand with those changes, specifically, the fairings necessary to channel air to the cattle in the trailers.

Meanwhile, engineers at Dryden did not simply file their reports and shelve the ideas they had been working on since the mid-70s. On the heels of the paper Saltzman, Wang, and Iliff presented at a 1999 AIAA meeting, Stephen A. Whitmore and Timothy R. Moes, both aeronautical engineers at the center, took up work on the idea, work that rolled into a new millennium. Using a small wind tunnel at Dryden, the engineers began looking at the drag minima that Saltzman had recognized following the Phoenix meeting. Motivating

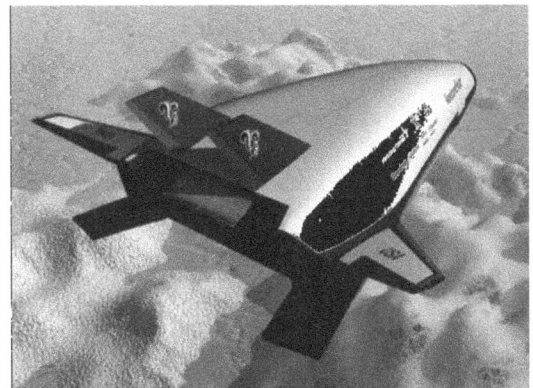

Lockheed's X-33 reusable space vehicle, like virtually all its family of predecessors, was a blunt aft body vehicle, a feature dictated by use of a rocket motor. In this case the vehicle was to be powered by a linear aerospike engine. The project never got off the ground.
NASA   EC99-44921-1

this particular research was not the trucking industry, however, but questions about shapes of vehicles entering the atmosphere from space. Coming full circle from the original experiments done by Saltzman and Horton, Whitmore and Moes completed studies on the drag characteristics of the X-33 and its linear aerospike engine, an experimental reusable spacecraft employing a radically different rocket motor.[23] As with other lifting bodies and space vehicles, this one had a flat back surface where the innovative rocket engine's nozzle sat. The engineers' objective was to see whether the overall drag of the X-33 could be improved with attention to the forebody.

A month after Whitmore and Moes published their results from the X-33 and LASRE (Linear Aerospike SR-71 Experiment) drag studies they'd conducted, in November 2000, Saltzman and three colleagues

---

[20] See, for example, three recent articles about trucking fleet acquisitions of features such as side skirts: "Utility Trailer side skirts receive SmartWay verification," *FleetOwner*, 18 February 2010, http://fleetowner.com/equipment/news/utility-trailer-side-skirts-0218/ (accessed 2 June 2010); Spirit Truck Lines of San Juan, Texas, said it ordered 300 trailers with Utility Trailer Manufacturing Co.'s newest trailer side skirts. The Utility Side Skirt 160 will cut Spirit Truck Lines' fuel consumption by 5%. "Fleets Install Aerodynamic Truck, Trailer Parts," *Transport Topics On Line*, March 29, 2010, http://www.ttnews.com/articles/basetemplate.aspx?storyid=24291 (accessed 2 June 2010); and "Wal-Mart May Save $300 Million with Fleet Efficiency," Clean Fleet Report, August 4, 2009, http://www.cleanfleetreport.com/fleets/wal-mart-to-save-300-million-with-hybrids/ (accessed 2 June 2010).

[21] "Cattle pot" is a slightly derisive term for a livestock trailer.

[22] Saltzman to Gelzer, notes on manuscript draft.

[23] Stephen A. Whitmore and Timothy R. Moes, *A Base Drag Reduction Experiment on the X-33 Linear Aerospike SR-71 (LASRE) Flight Program* (Edwards, CA: NASA TM 1999-206575, 2000).

62 **Results**

The Linear Aerospike SR-71 Experiment (LASRE) involved a scaled linear aerospike engine mounted on the dorsal area of a NASA SR-71.
NASA   *EC96-43419-19*

proposed a new research project to further explore what Moes and Whitmore were looking at and what Saltzman had first remarked on after the Phoenix meeting. The plan was to, once again, use a ground research vehicle. "The optimum forebody drag is not necessarily obtained by minimizing the forebody drag," noted the four Dryden researchers in their proposal. "In fact, in the sub-optimal region, as forebody drag is increased, base drag will decrease causing an overall reduction in vehicle drag [emphasis added]."[24]

Center managers approved the research program, which led to a new ground research vehicle affectionately dubbed the Roadrunner. This one was a Chevrolet heavy-duty van that was once again modified by the center's motor pool and fabrication shops. Whereas only the exterior of the Shoebox had been modified, the Roadrunner was altered on the inside as well. Shop technicians cut apart the vehicle's frame and fabricated additional framing that allowed the crew to open the vehicle and add a segment that would lengthen the machine, or remove it and shorten the vehicle.[25] At its greatest length the Roadrunner had a fineness ratio representative of a typical tractor-trailer combination. Technicians made it possible to disengage the driveline in order to reduce the mechanical drag during coast-down tests. Armed with Saltzman's

---

[24] Tim Moes, Tony Whitmore, Ken Iliff, and Ed Saltzman, "Base Drag Research Using a Ground Research Vehicle," internal memo seeking funding approval for a new project, 12 October 2000, 1.

[25] Increasing the vehicle's length relative to its diameter was essential to the engineers, who wanted to factor Reynolds numbers in to their investigations, something that changing both the vehicle's fineness ratio and surface roughness made possible.

In 2000, NASA Dryden began a research project with a new ground vehicle dubbed the Roadrunner. The fabrication shop cut off the rear portion of a Chevrolet van (at left, top and bottom) and cut the chassis in half so a section could be added to lengthen the vehicle in order to change achieve different Reynolds numbers (above).
Clockwise: NASA EC99-45167-01, EC99-45167-03, EC99-45167-16

Here, the Roadrunner's substructure is visible as it is being built up. Noticeable is that the framework allows the back section to be removed altogether, changing the Reynolds numbers of the vehicle in motion.
NASA   EC00-0116-01

revelation regarding finite returns on aerodynamic modifications for blunt vehicles, Corey Diebler and Mark Smith undertook tests using the Roadrunner. They set out to evaluate Hoerner's original two- and three-dimensional equations for base drag predictions, starting with the hypothesis that "a new base drag prediction model [was] needed for large-scale vehicles."[26] For while Hoerner predicted a rise in base drag as forebody drag was reduced, research on the Shoebox, reinforced by data collected with blunt air vehicles, showed that "Hoerner's formula greatly underestimates this dependence of base drag on forebody efficiency."[27]

Following the initial baseline tests, they moved to the next phase, starting with a deceptively simple idea: that coarsening the vehicle's forebody surface roughness would thicken the boundary layer, which in turn

---

[26] Diebler and Smith, A Ground-Based Research Vehicle, 12.

[27] Edwin J. Saltzman and Robert R. Meyer Jr., A Reassessment of Heavy-Duty Truck Aerodynamic Design Features and Priorities (NASA/TP-1999-206574, June 1999), 3. "The demonstrated increase in base drag associated with forebody refinement indicates that the goal of a 0.25 drag coefficient will not be achieved without also reducing afterbody drag. A third configuration of the test van had a truncated boattail to reduce afterbody drag and achieved a drag coefficient of 0.242."

64 Results

Above and below, the Ground Research Vehicle (Roadrunner) during runs on the NASA Dryden flight line. Tufts of yarn show chaotic airflow over the vehicle's first half but fairly smooth and controlled airflow over the latter half. A rake, used to measure boundary layers, stands above the roofline at the vehicle's aft end.
NASA *EC02-0208-02, EC02-0208-05*

Members of the fabrication shop team responsible for the Roadrunner's structural modification. From left are Tom King, Jason Denman, Jerry Reedy, and Dale Hilliard.
NASA   *EC00-0016-09*

would reduce the pumping action of the airflow as it encountered the base of the vehicle. A thicker boundary layer generates more drag, but this is a minor penalty compared to the reduction in pressure drop at the vehicle's aft end and the ensuing drag reduction—up to a critical point, of course. "The phenomena [drag bucket] is directly related to the boundary layer thickness," noted Moes, Whitmore, Illif, and Saltzman in their proposal. As airflow becomes laminar it loses energy; this, in turn, makes it increasingly less likely that this flow will remain attached to the surface over which it is flowing. In contrast, a turbulent boundary layer energizes the airflow, helping to keep that airflow attached to the vehicle further along its surface (wing or trailer). Though a turbulent boundary layer also adds some drag as compared with laminar flow, the payback comes by keeping that airflow attached to the surface longer, thus reducing a dramatic increase in drag. Driving the new research was the counterintuitive fact that is an attribute of the drag bucket: "For certain configurations," read an abstract from a paper written early in the project, "the total drag of a vehicle can be reduced by increasing its forebody drag" [28] The research would be applicable on such things as trucks, buses, and motor homes, not to mention reentry vehicles.

---

[28] Corey Diebler and Mark Smith, *A Ground-Based Research Vehicle for Base Drag Studies at Subsonic Speeds* (Edwards, CA: NASA TM 2002-210737, November 2002), 1.

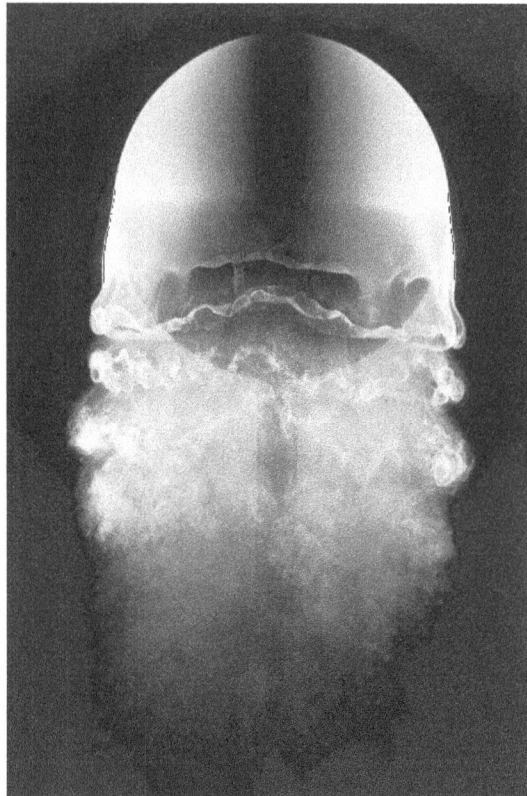 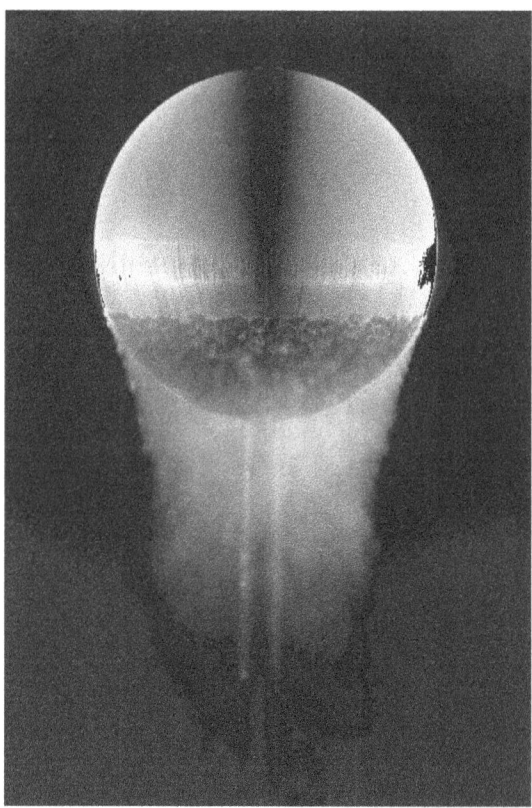

Roughening a surface can reduce drag, something baseball pitchers and golfers know well. In the image at left, a laminar boundary layer begins to separate just ahead of the sphere's "equator;" it soon breaks away from the sphere entirely and becomes turbulent. At right, a wire at the "equator" trips the boundary layer, initiating turbulence. As a result, the flow adheres to the surface farther downstream.
Images by Henri Werlé, courtesy Onera, the French Aerospace Laboratory. C6976, C6980

Researchers examining fluid dynamics and golf ball makers knew the counterintuitive: the best way to reduce drag on a laminar flow object was to increase drag on the forebody. Adding drag by creating an energized airflow seems at odds with the truck fairing program's objectives, but doing so paid higher dividends than the penalty of added forebody drag.[29] The critical element in this is that the base drag does not drop as much with a thick boundary layer as it does with a thin layer, and the objective is to raise base pressure in order to reduce aerodynamic drag. As with many projects, the Roadrunner remains unfinished. Larger imperatives drew the researchers away to other projects and did not allow them to return as they would have liked.

But unfinished does not mean it's over, as the Phoenix meeting proved for Saltzman and the "drag bucket," so opportunities remain to expand on the foundation built here at the NASA center in the desert. As fuel costs continue to climb, there remains the question of whether long-haul trucks lacking aerodynamic improvements will continue to ply the nation's roads. The same question can be asked of the motor home industry, which seems on the surface to be largely immune to concerns about vehicle aerodynamics. Concessions to ideal shapes over functionality appear necessary before motor homes exhibit the changes tractors—and some trailers—have undergone.

---

[29] "A turbulent boundary layer that is thin is an efficient pump at the aft edge of the vehicle; it sucks down the base pressure to a low value and increases base drag. A thick boundary layer remains attached to the surface, but because it is thick it is a less efficient pump and cannot pump down the pressure on the base surface as low as the thin boundary layer can." Saltzman to Gelzer, notes on manuscript draft.

Chapter Eleven
The Social Construction of a Technology

The changes discussed here are physical ones; the intangible is the cultural change that accompanies the material reshaping of our world and things in it, and the initial reaction to the FRC highly faired tractor-trailer unit and to the Kenworth T600 suggests more than a few questions. For instance, if making money is central to American culture, then making that money more efficiently would seem a socially acceptable pursuit. But is it?

The small population of colonists in North America—small relative to the continent's landmass when compared to that of Europe—birthed a trend for efficiency and speed starting in the eighteenth century that many in Europe remarked upon.[1] Take pelagic whaling, for instance, which was a European invention. A ship put to sea, its sailors killed and butchered whales then stored the fat in hogsheads below deck until all were full, and then the ship returned to port, where the blubber was cooked down into oil in try-works. So long as whaling was confined to the polar regions things were not so bad. But as whalers wandered further south in search of prey the blubber decayed faster, making whaling ships of the seventeenth and eighteenth centuries vessels to be avoided at all costs.[2] An American captain is credited with first putting a try-works on board a whaler in 1715: Christopher Hussey chose to take the factory to sea with him in the form of two large cauldrons held in place above a brick structure. Whale blubber was rendered into oil in the cauldrons and then stowed below, where it had a nearly indefinite life—and no particular odor. Stowing oil rather than blubber was a far more efficient use of limited space, as well. Once started, the try-works was fed with cracklings from the cauldrons, making the process almost self-sustaining.[3] It wasn't long before American whalers adopted this method en masse and soon dominated whaling the world over.

And whaling wasn't the only example of a quest for greater efficiency (and so, a better return on an investment). In 1782 Oliver Evans built what is considered to be the first automatic flourmill, in Delaware. While elsewhere everything but grinding the grain was done by hand, Evan's Red Clay Creek Mill automated each step of the process with machinery, powered—like the

---

[1] See, for example, Brooke Hindle and Steven Lubar, *Engines of Change: The American Industrial Revolution, 1790-1860* (Washington, D.C.: Smithsonian Institution Press, 1986), as just one book on the subject. Observed Englishman J. Milton Mackie in the early 1860s during his tour of the U.S.: "Every man is either just in from Cincinnati or Chicago, or he is starting for one of these places. Unless he makes his hundred miles between breakfast and dinner, he counts himself an idler and talks of growing rusty. I saw in the West no signs of quiet enjoyment of life, but only a haste to get rich. Here are no idlers." J. Milton Mackie, *From Cape Cod to Dixie and the Tropics*, 2 volumes (New York: Putnam, 1864), 192. Ebenezer Davies, a fellow English tourist, was asked to preach a sermon while visiting Cincinnati. Of the experience he later wrote: "At the close of the sermon, having pronounced the benediction, I engaged, according to English custom, in a short act of private devotion. When I raised my head and opened my eyes, the very last man of the congregation was actually making his exit through the doorway; and it was quite as much as I could manage to put on my top-coat and gloves and reach the door before the sexton closed it. This rushing habit in the House of God strikes a stranger as rude and irreverent. A man marches into his pew, or his pulpit, sits down, wipes his nose, and stares all about him; and at the close, the moment the 'Amen' is uttered he is off with as much speed as if his house were on fire." Ebenezer Davies, *American Scenes, and Christian Slavery: A recent Tour of Four Thousand Miles in the United States*. (London: John Snow, 1849), 151.

[2] There are accounts from this period of whale blubber fizzing in the barrels below decks as it decayed while the whaler sailed on for more cargo, and the description of the stench that clung to these vessels leaves little to the imagination.

[3] Samuel Elliot Morrison, *The Maritime History of Massachusetts, 1783-1860* (Boston: Houghton Mifflin, 1961), 19-21.

grist mill itself—by water, so that what emerged at the end was bagged flour that had merely to be sewn closed.[4] Or take Thomas Blanchard, who, by 1825, had created a set of machines that became central to gunmaking in U.S. armories. A particularly ingenious invention of his was a lathe capable of carving irregular shapes, such as gunstocks; heretofore, lathes were only useful in turning symmetrical items. The result was that it now cost less to make a rifle because unskilled labor (even boys) could tend the lathes, and it took far less time to make them as well. The ability to make genuinely interchangeable parts, something often (but incorrectly) credited to Eli Whitney, also has its roots in the U.S. armories. As machines replaced gunsmiths—and produced more parts that were genuinely interchangeable, in the process—those machines drove down the cost of guns while increasing production. Not surprisingly, the gunsmiths reacted negatively, but neither Evans nor Blanchard nor anyone else associated with what eventually came to be known as the 'American system' suffered general public scorn for improving an existing system or seeking efficiency.[5] Rather, people like Cyrus McCormick (who did not invent the mechanical reaper but did figure out how to mass produce and especially market the machine effectively), went on to be celebrated for their contributions.

What these examples epitomize is a quest for efficiency, typically in manufacturing, which drew the attention (and frequently the admiration) of many in the world. That efficiency, in turn, stood to improve the owner or operator's standing in the community by increasing his wealth, which in turn was the principal means of measuring worth in a society that rejected an inherited aristocracy. Writing about the nineteenth century, historian John Kasson noted that "Americans' intense aesthetic response to technology and their desire to discover beauty in utility were rooted in republican values," since this technology provided opportunity for individual advancement.[6] Frenchman Alexis de Tocqueville, ostensibly sent by his government in 1831 to survey the American penal system, spent two years touring the United States. Once back home he wrote what is still considered one of the best and most comprehensive early assessments of American (U.S., not Colonial) culture by a foreigner, which he subsequently published in two volumes. Americans, he said, "cultivate those arts which help to make life comfortable rather than those which adorn it. They habitually put use before beauty, and they want beauty to be useful."[7]

Yet if we move ahead more than 200 years we find that when it came to trucking, earning more than your neighbor by driving a more efficient truck clearly carried with it unmanly overtones, as references to the

---

[4] "The Genius of Oliver Evans," Joseph Gies, *AmericanHeritage.com*, http://www.americanheritage.com/articles/magazine/it/1990/2/1990_2_50.shtml (accessed 3 June 2010). Evans received U.S. patent no. 3 for his mill. He also designed and built the *Orukter Amphibolos*, a steam-powered, self-propelled amphibious dredge for the city of Philadelphia, in 1805. He drove the 17-ton machine awkwardly through the streets and into the water on its own power, and though it was a remarkable idea, it was ahead of its time and failed to meet expectations. While others relied on relatively low-pressure steam engines for power, Evans favored high-pressure steam, and designed and built his own engines. His steam engines were smaller, lighter, and more powerful as a result.

[5] Hindle and Lubar, *Engines of Change*, 225-233. Blanchard's first successful use of his lathe to cut irregular shapes was with shoe lasts. Regarding changes in the armories, see Merritt Roe Smith, Harpers *Ferry Armory and the New Technology: The Challenge of Change* (Ithaca: Cornell University Press, 1977). Not surprisingly, the arsenal craftsmen resisted the changes brought by industrialization and a violent strike ensued. Worse, when superintendant Thomas Dunn enforced new work patterns, a disgruntled one-time employee murdered him. In the end, however, the new work patterns forced out both the craftsmen and their ways. The matter of efficiency in the workplace is not nearly as simple as this short version would suggest, of course. It involves putting "skill" in a machine that, in turn, makes labor generally the unskilled partner in the process, a shift in control of the workplace from worker to manager, and a host of benefits and tensions, to list but three elements. In the case of Harpers Ferry, the community itself was central to the story itself, as well. The compulsion for efficiency acquired new momentum in the early twentieth century with the advent of Taylorism and Scientific Management, and efficiency remains a driving force in modern society.

[6] The republican values Kasson referred to were not those of a political party but those of a political philosophy of individual rights and liberty, with a government responsive to its citizenry, and which was in marked contrast to a European tradition of hereditary nobility. John Kasson, *Civilizing the Machine: Technology and Republican Values in American, 1776-1900* (New York: Grossman Publishers, 1976; reprint, New York: Penguin Books, 1977), 144.

[7] Alexis de Tocqueville, *Democracy in America*, translated by George Lawrence and edited by J. P. Meyer, vol. 1, (New York: Harper & Row; reprint by Anchor Books, 1969), 465. See also Zoltan Simon, *The Double-Edged Sword: The Technological Sublime in American Novels, 1900-1940* (Budapest: Akademia Kiado, 2003).

center's "sissy truck" indicate. And when Contract Freighters began buying the T600, the truck brought some men to tears at the thought of being seen driving what they called "the anteater," and some men literally pleaded not to have to drive one because of the personal embarrassment it would mean. If the comparison between whaling or milling and trucking is simplifying the matter, the larger point remains: being efficient at a task can have, at least for some Americans and in certain contexts, effeminate overtones.[8] While the "sissy truck" reaction is not as pervasive in 2010 as it was in 1975, that sentiment still echoes in the present century: consider that Bob Weber explained the enduring non-aerodynamic designs by referring to some drivers as "pride & polish . . . hardcore image guys." The Mack Company's "Rawhide" model was named to appeal to a particular segment of the market that was one once filled by trucks deemed "tough-looking, brawny, hairy-chested."[9] The Rawhide's features intentionally made no concessions to aerodynamics. Plainly, there is a segment of the market principally comprising owner-operators who willingly sacrifice income for appearance, and that segment remains large enough—even in the twenty-first century—that truck manufacturers find it economically viable to cater to it. Is there something manly in squandering money? Conversely, is it effeminate to be efficient?[10]

Or is it a matter of the suspicion with which Americans have long held clever, mental work and the warm regard they have long had for the fruits of hard labor and the physical challenges of life? Americans have a centuries-old ambivalence bordering on outright distrust for the intellectual approach to things, as Richard Hofstadter so deftly pointed out.[11] In the 1827 presidential campaign between John Quincy Adams and Andrew Jackson, for example, the following ditty

---

[8] This is quite likely a function of a more fluid society in terms of both work and gender roles compared to three centuries earlier, leaving it to individuals, rather than their work, to define who they are.

[9] Schenck, "New Focus on Air Drag," 37. The description referred to boxy COEs in contrast to the relatively curvy 1995 GM Astro COE that was, itself, not particularly aerodynamic.

[10] One of the more accessible discussions on the topic of wealthy individuals choosing to squander wealth with social objectives as their motivation can be found in Thorstein Veblen's *The Theory of the Leisure Class: An Economic Study of Institutions* with a Foreword by Stuart Chase (New York: The Modern Library, 1899; reprint, 1934).

Economists sometimes invoke the term "signaling" when discussing efforts by one group or an individual to transmit a particular message to others, albeit obliquely. One of the most identifiable examples of this was the Space Race of the 1960s, in which the Soviets and Americans sought to convince much of the Second and Third worlds of the super powers' respective superior qualities in the hopes of enticing undecided nations to align with one or the other on the global stage. It wasn't merely the extraordinary feat of going into space that the two nations put up for people to judge, but the staggering cost involved, which no other nation could incur—and everyone knew it. An essential element of signaling, therefore, is economic sacrifice. Why? In addition to separating people or groups into discernable categories (I can, you can't), the heavy cost may be all that distinguishes the genuine signal from a fake. Owner-operators who consciously buy a less efficient truck—one that is visibly inefficient, it is important to note—are, themselves, signaling something; just what that might be is the subject of another study. A fine discussion of signaling can be found in Alexander MacDonald, "The Long Space Age: Essays on the Economic History of Space Exploration from Galileo to Gagarin," D.Phil. diss., Baliol College, Oxford University, 2010.

[11] Richard Hofstadter, *Anti-Intellectualism in American Life* (New York: Alfred A. Knopf, 1963, 5th edition; reprint, 1969), which ironically won a Pulitzer Prize. In *The Self Made Man In America: The Myth of Rags to Riches*, Irving Wyllie notes that throughout the nineteenth century orators and authors spoke highly of mediocrity and seemed to look askance at genius. The latter was disparaged for relying on innate ability to achieve things, the former lauded for perseverance. Or, as Richard Huber put it: "All success writers of the character ethic believed in the value of self-education. There was considerably less unanimity on the value of formal schooling, and even less on the necessity of it." Irving G. Wyllie, *The Self-Made Man in America: The Myth of Rags to Riches* (New York: The Free Press, 1966), 95. It did not help dispel the notion in the long run that a study done by Pitirim Sorokin in 1925 concluded that 88.3 percent of the American millionaires who rose from "humble" beginnings had no more than a high school education, and that 71.7 percent of these had a grade school education, at best. To these heroes a classical education was worth little, and many of them railed against the waste of time that such effete learning entailed. Of course, collegiate and university education was not nearly as accessible then as it is now. Pitirim Sorokin, "American Millionaires," *Journal of Social Forces* 3 (1925): 637. Richard M. Huber, *The American Idea of Success* (New York: McGraw-Hill Book Company, 1971), 161.

was quite popular:

> John Quincy Adams who can write
> And Andrew Jackson who can fight[12]

Jackson won two consecutive terms.

Although education has taken on greater esteem in American society since then, popular culture remains a good measure of what career paths are valued most: even today, sports figures are far more prominent as icons than rocket engineers or public servants. Indeed, college dropouts who then succeed in business, though infinitesimally small in number when compared with the number of individuals at work in the larger corporate world, continue to capture the popular imagination if for no other reason than because of what they accomplish despite lacking a customary education. Nineteenth-century steamship and railroad baron Cornelius Vanderbilt was proud of having had no schooling to speak of, and made no apologies for his utter disdain for books. Late in life Andrew Carnegie endowed libraries around the country, but he was quite proud of having made his fortune as a young man without any formal education, relying instead on his hands, initiative, and wits; "dead languages," Carnegie called classical education. More recently, Microsoft's Bill Gates achieved business and financial success despite being a college dropout, as did fellow computer mogul Steve Jobs of Apple, who abandoned college after one semester; both apparently demonstrated the triumph of common sense over sound education.

How then to explain the initial reaction among drivers to the FRC's highly faired truck, or the T600 that drivers dubbed the "anteater"? And how to explain the persistence of new, non-aerodynamic tractors, whose popularity is sufficient to warrant continued manufacturers' investment? In this vein, what should we make of choices about gendered artifacts and technologies in instances where there is a clear penalty for choosing the less efficient or less popular option? Despite the context in which it is posed, this is not a question with only negative answers. Is pride in work wrong, even if it costs more? Should efficiency, by whatever definition, be the only measure by which we determine value or worth?

And what should we make of choices about gendered careers? This is not an abstract question, either for engineering in general or for NASA in particular; the engineering profession has been and continues to be dominated numerically and in virtually every other way by men long after the disparity was recognized and schools made efforts to address it. The Federal Bureau of Labor Statistics found in 2009 that for every female in the fields of engineering and architecture there were 6.27 males.[13] That ratio, at 1:6.4, was not materially different from the rate in 2008. And yet some women continue to choose engineering as a profession, in spite of what may be a Sisyphean struggle.

The change in the shape of long-haul trucks, wrought in no small part by NASA engineers, has been extraordinary. Chiffrephiles could likely calculate how many gallons of fuel have been saved by the aerodynamic changes adopted since 1975. Whatever the reckoning, it would be impressive.[14] Just as interesting, however, are the meanings with which Americans invest their technology, because these have a direct impact on what Americans create, grant access to, embrace, or denigrate, or even choose to use.

---

[12] This sort of thing didn't start with Jackson. As early as the election of 1796, William Laughton Smith portrayed Thomas Jefferson as a philosopher in an era that needed a man of action, suggesting that Jefferson was not the right sort of person for the job of president: philosophers were apt to act on principle and be entirely too inclined to reason. What was needed, wrote Smith, was a man who understood action, the material world, and the demands of living rather than moralizing and thinking. William Laughton Smith, *The Pretensions of Thomas Jefferson to the Presidency Examined* (n.p., 1796) in Hofstadter, *Anti-intellectualism in American Life*, 14.

[13] Labor Force Statistics from the Current Population Survey. Bureau of Labor Statistics, http://www.bls.gov/cps/tables.htm (accessed 3 August 2010).

[14] Angus Maddison may not have coined the term chiffrephile but he happily considered himself one: a lover of numbers, not simply for numbers' sake but to make sense of things. Chiffrephiles go well beyond the typical economic theorist; Maddison's final intellectual adventure was to calculate the world's economic output in the year 1 A.D., which he figured to have been $105.4 billion in 1990 dollars. "Maddison Counting: A Long Passionate Affair with Numbers Has Finally Come to An End," *The Economist*, April 29, 2010, http://www.economist.com/research/articlesBySubject/displaystory.cfm?subjectid=348996&story_id=16004937 (accessed 1 June 2010).

Appendices

Appendix A

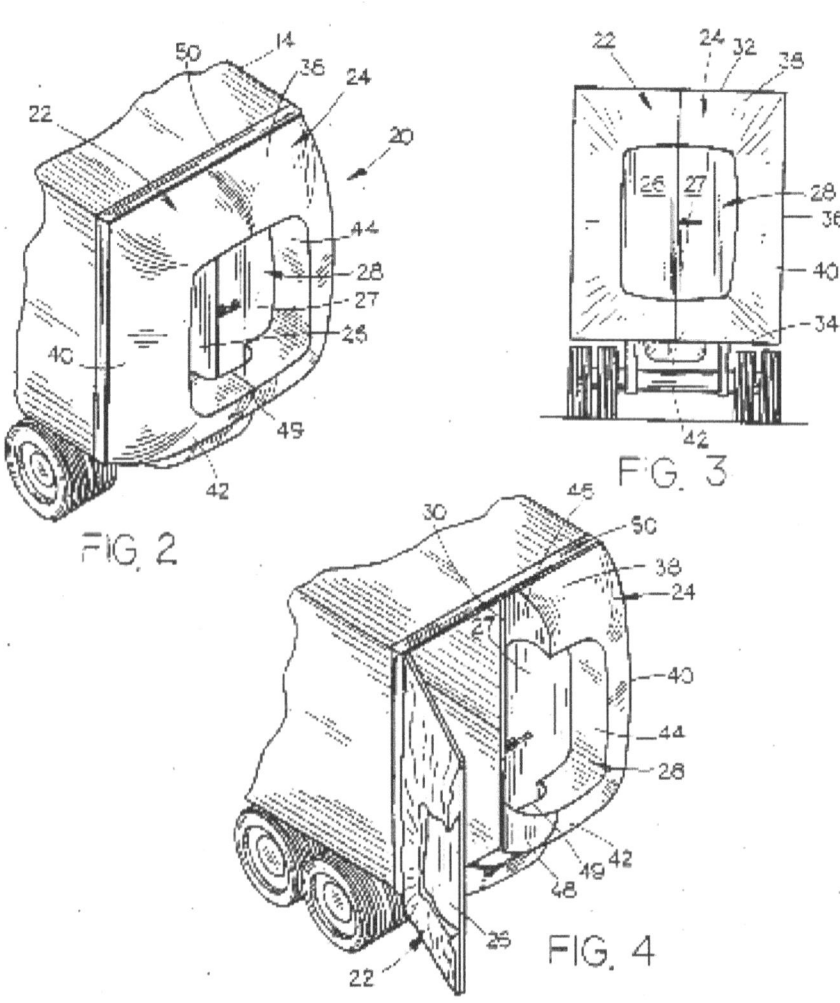

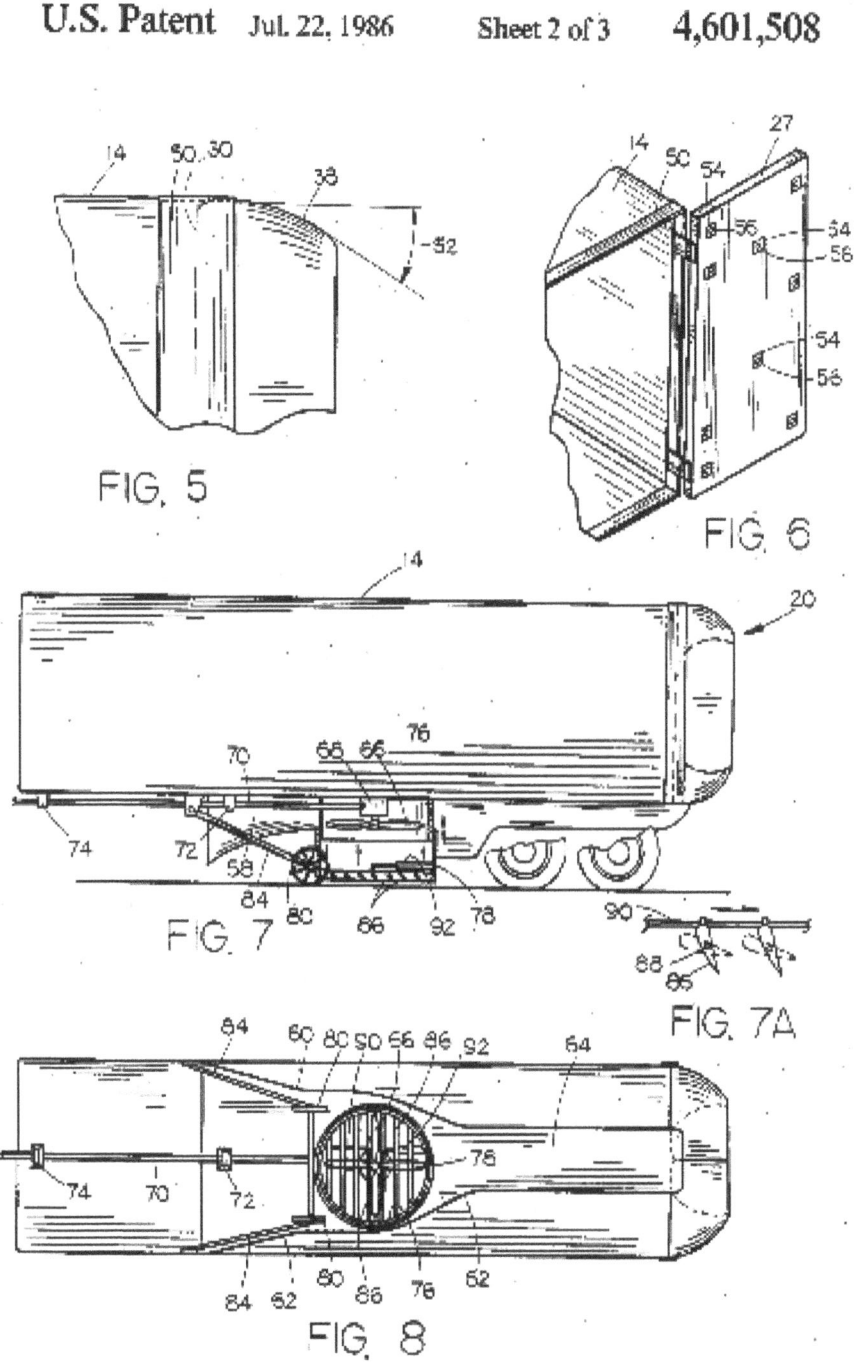

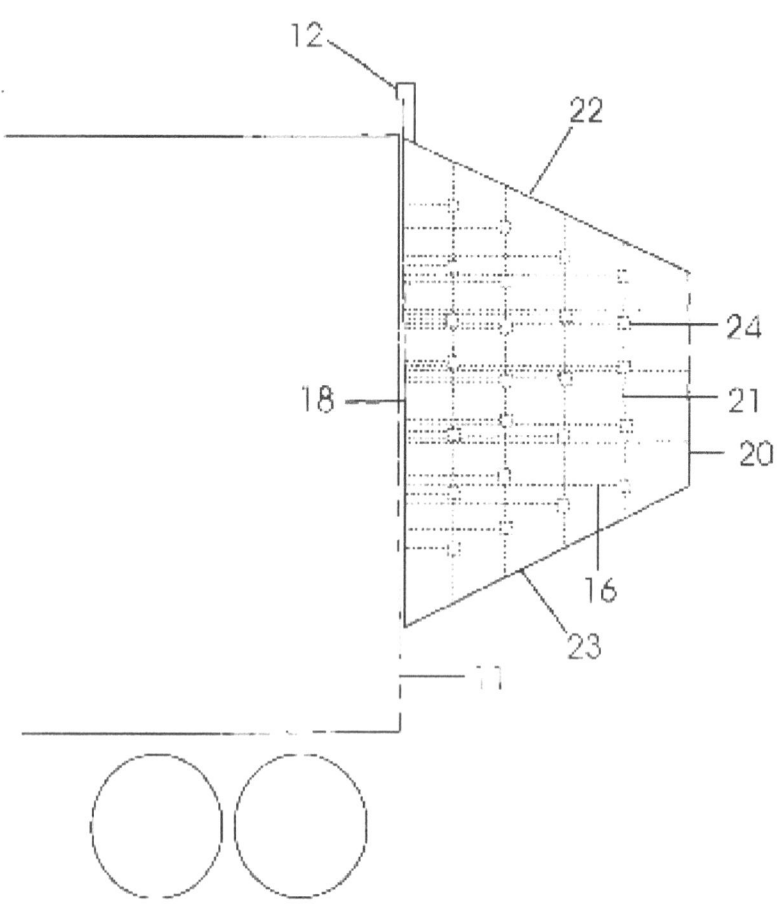

Figure 1

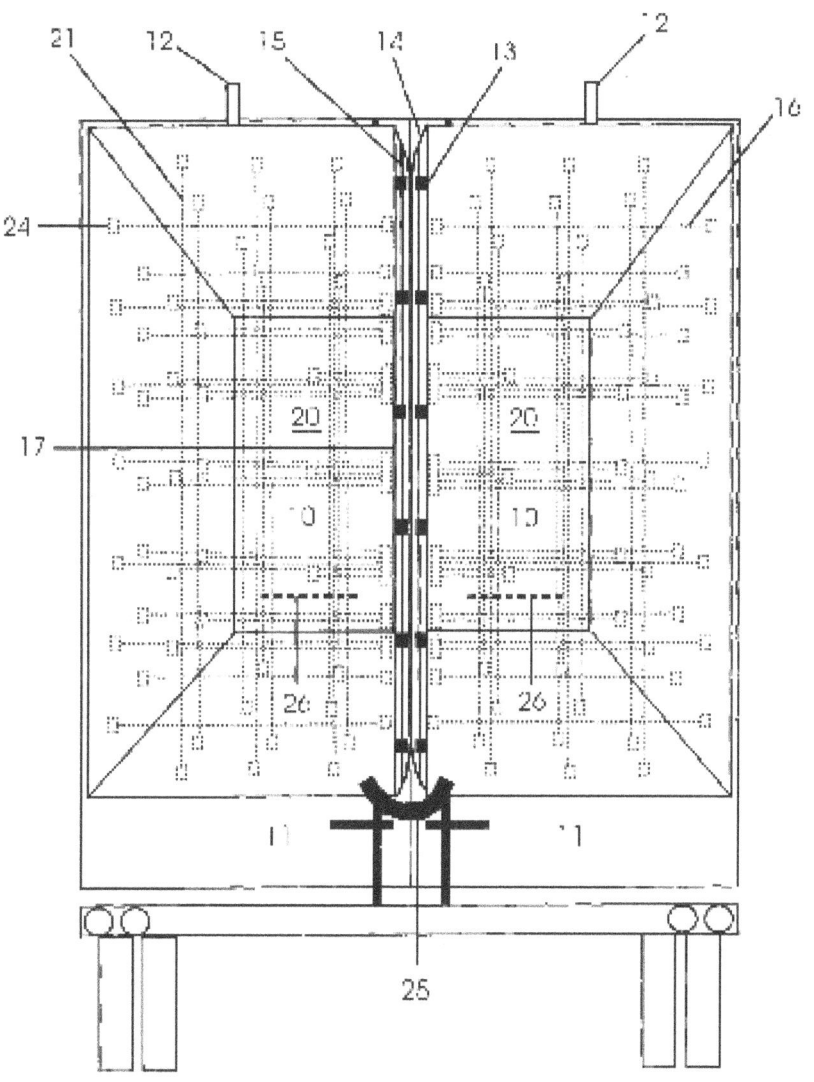

Figure 2

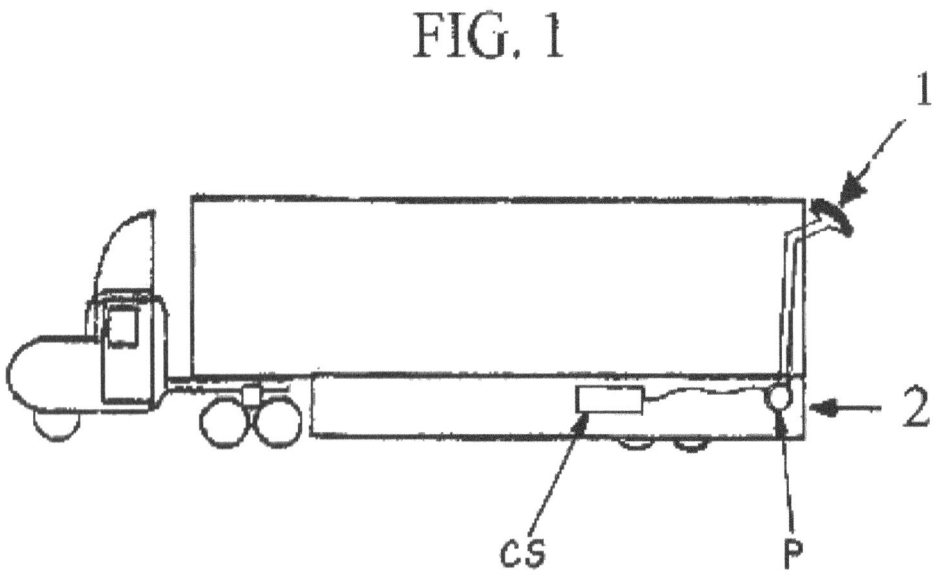

FIG. 1

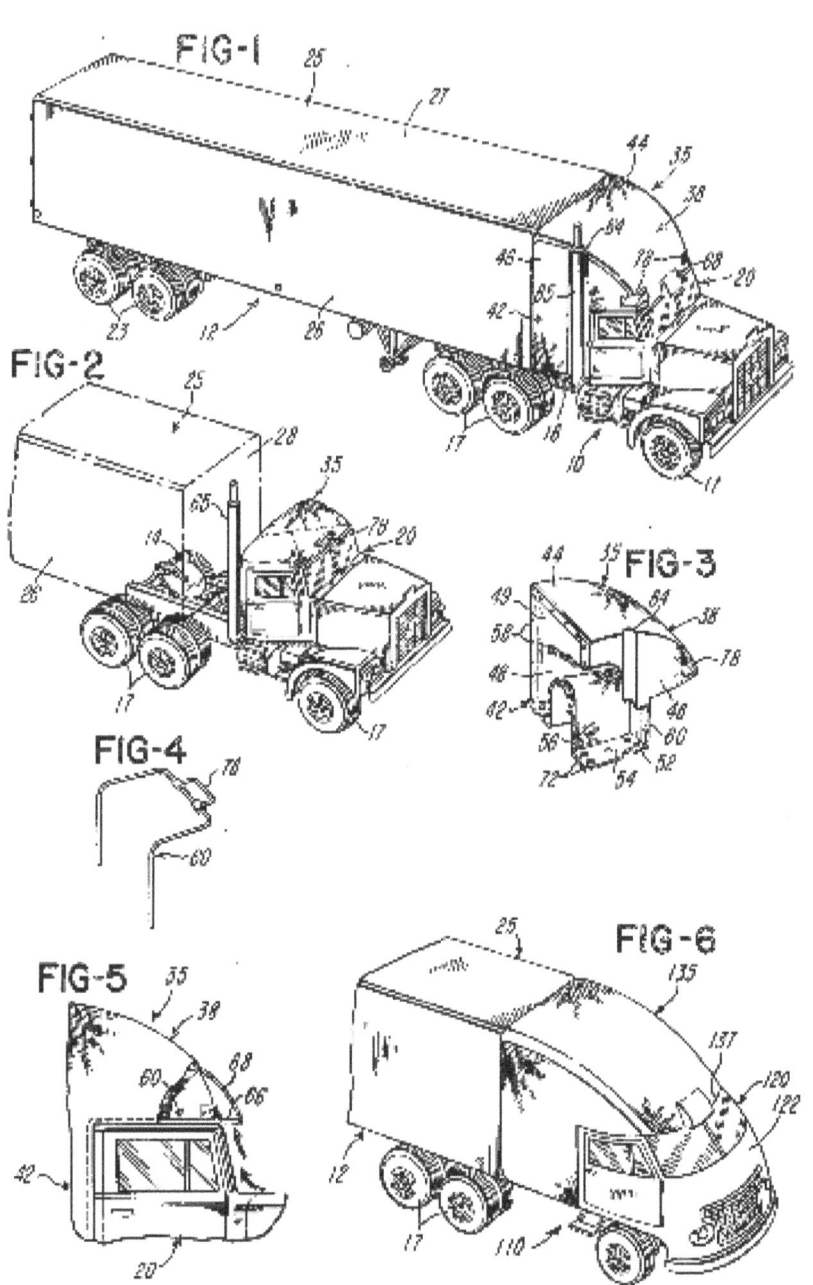

Appendices

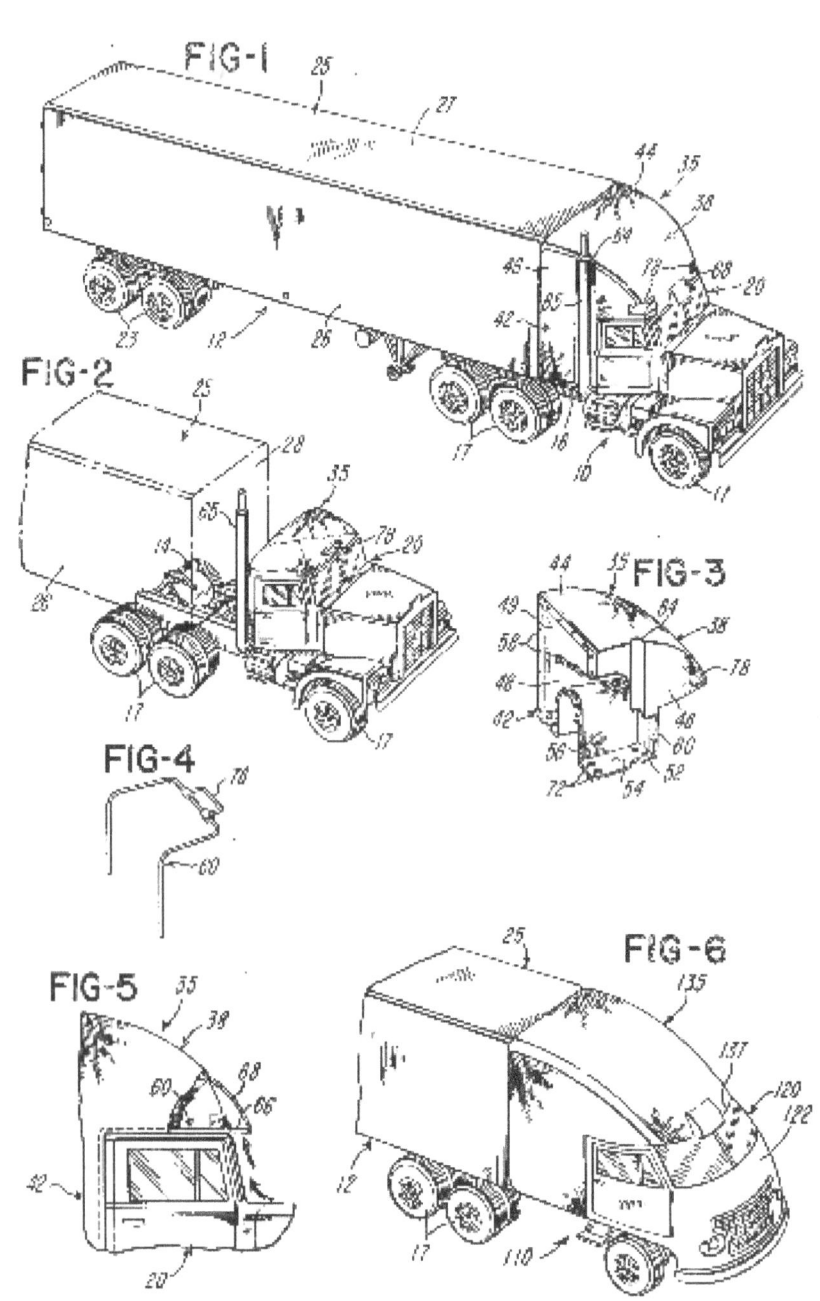

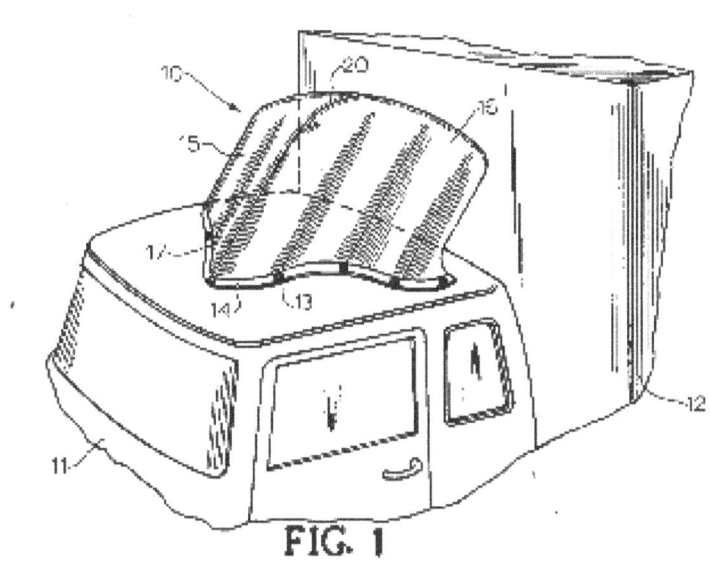

FIG. 1

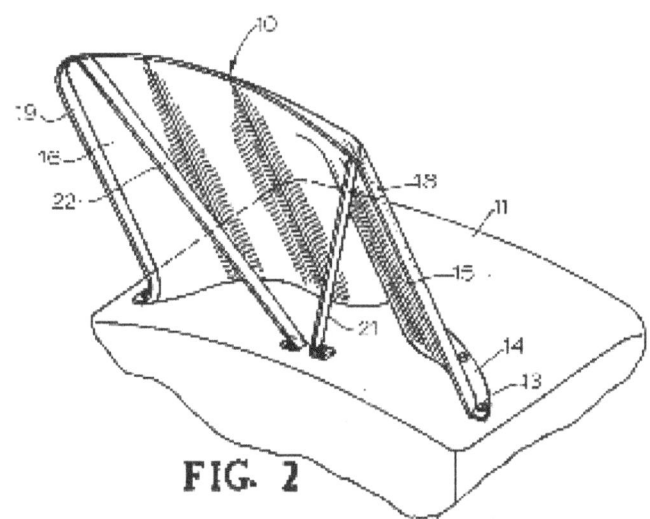

FIG. 2

Appendices

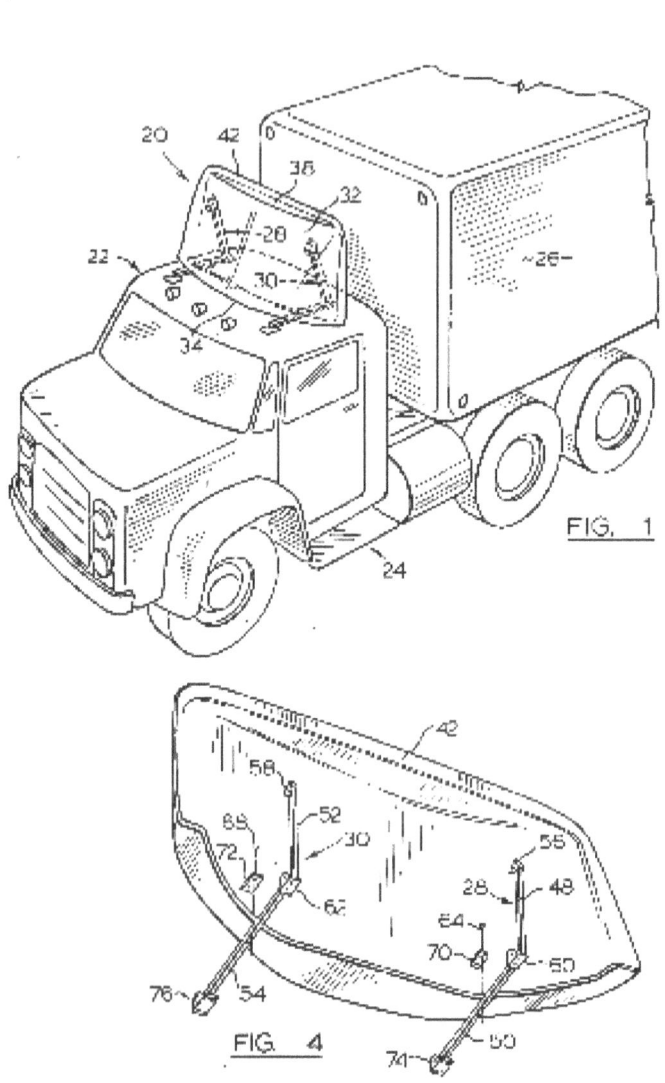

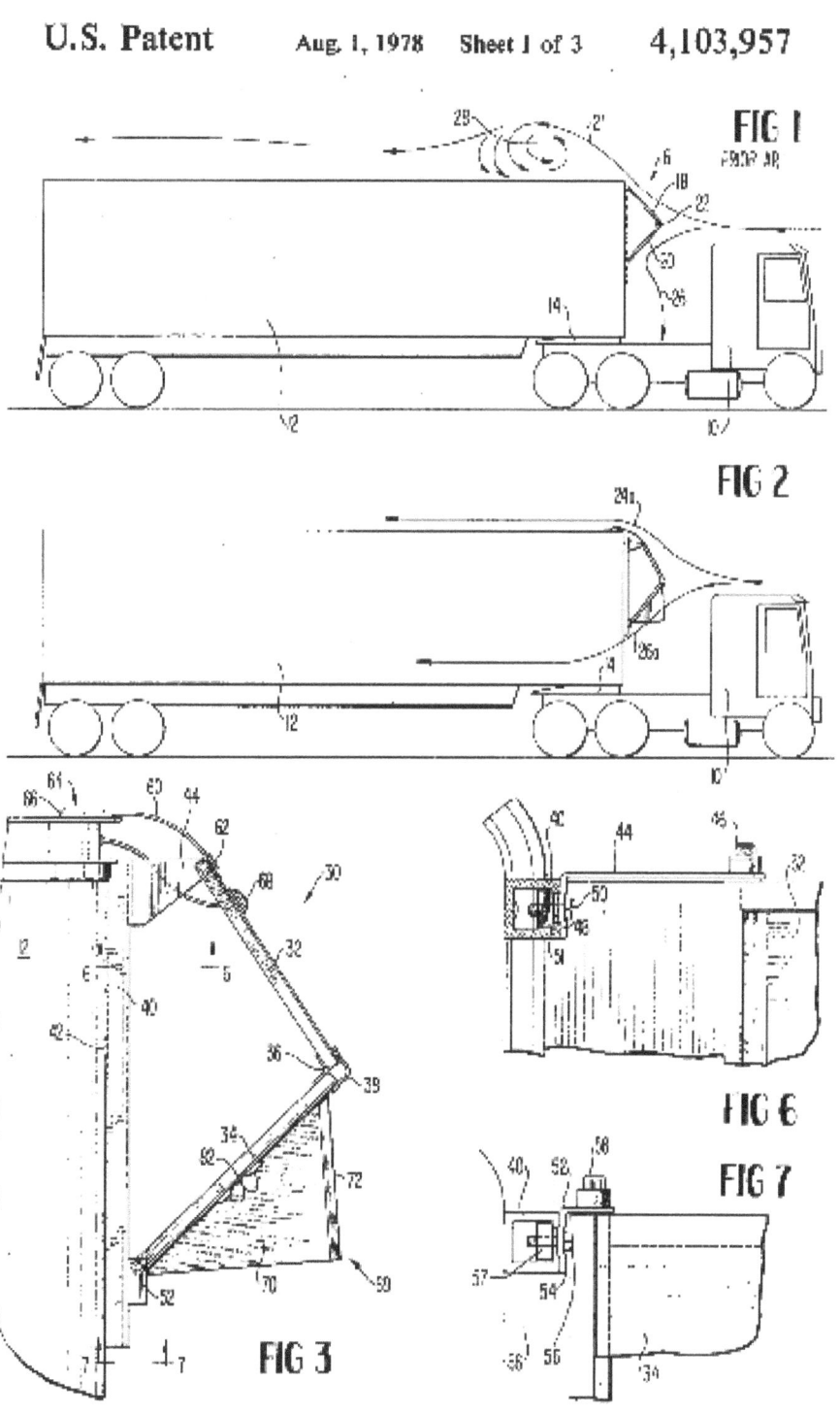

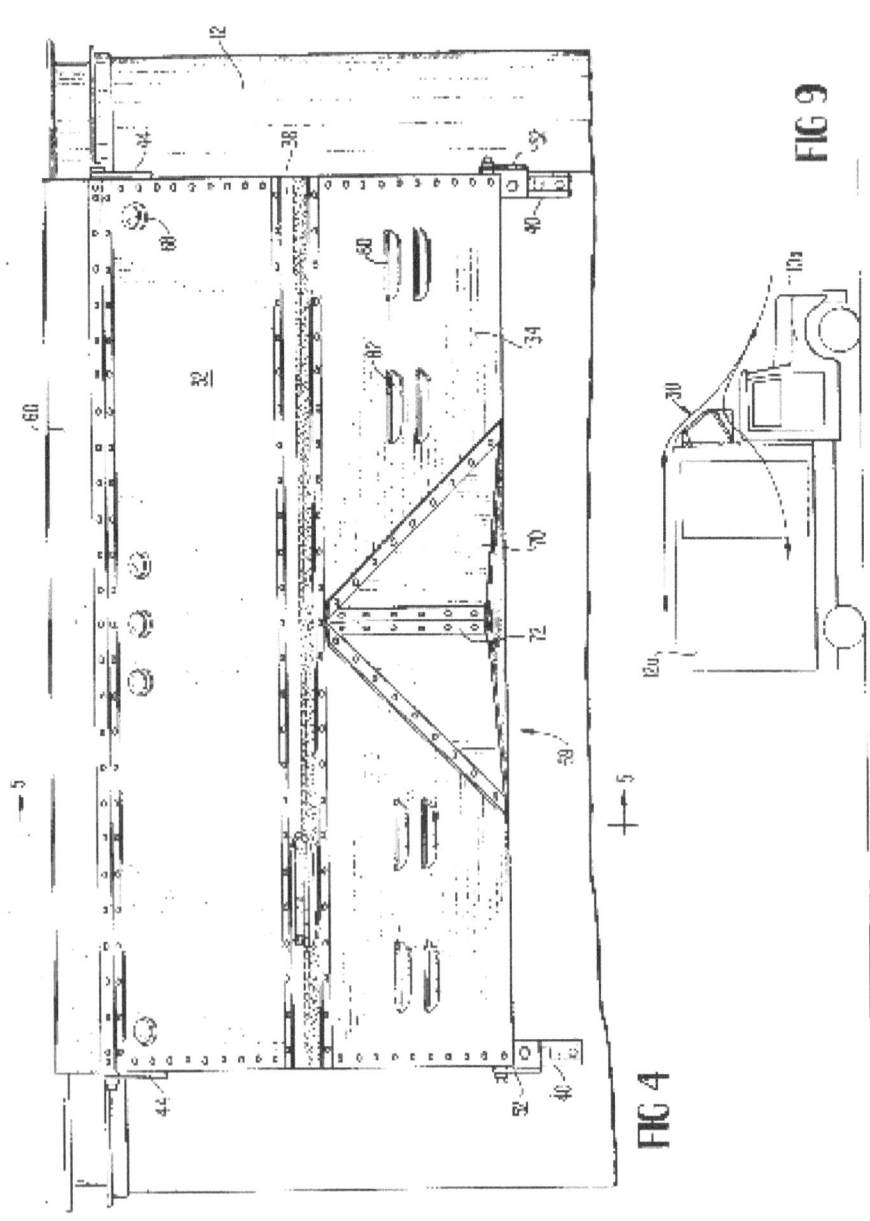

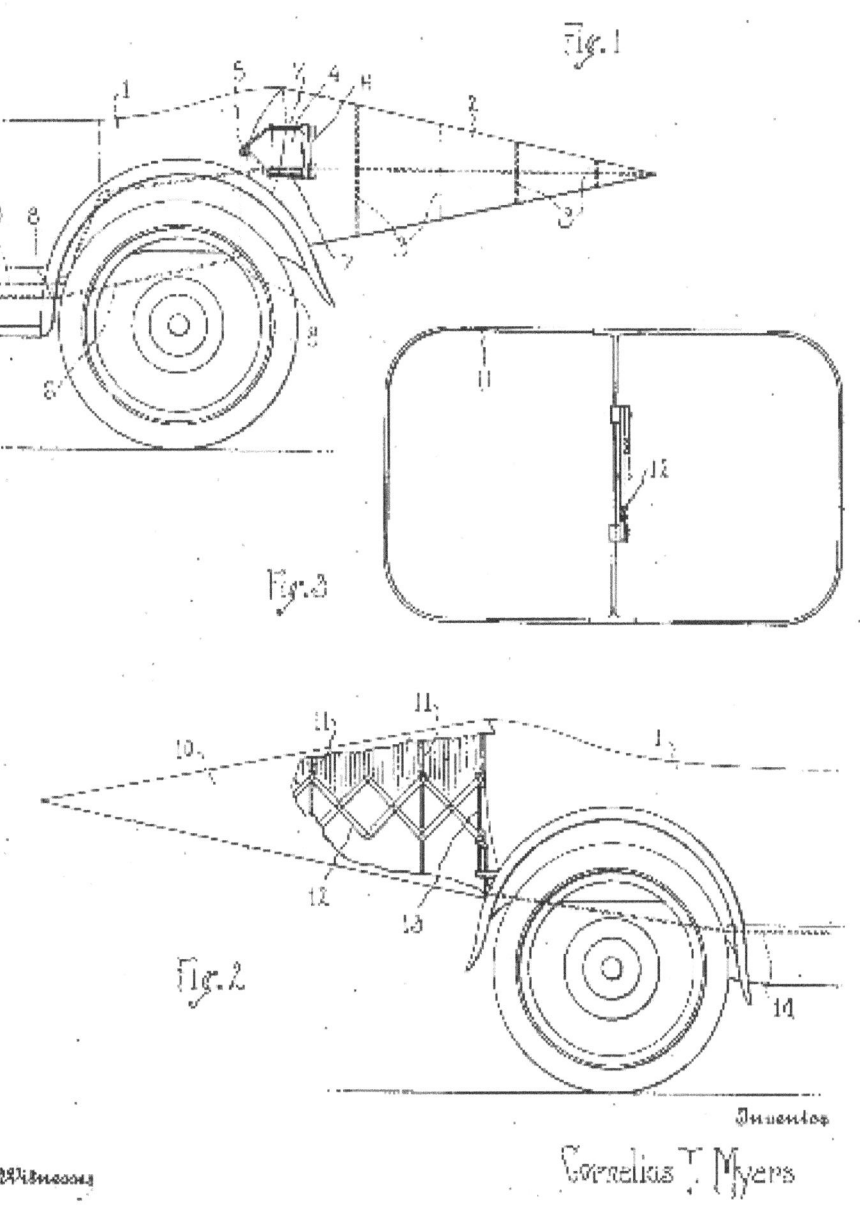

Appendices

April 21, 1936. E. A. STALKER 2,037,942
MEANS OF REDUCING THE FLUID RESISTANCE OF PROPELLED VEHICLES
Filed Oct. 28, 1935 2 Sheets-Sheet 1

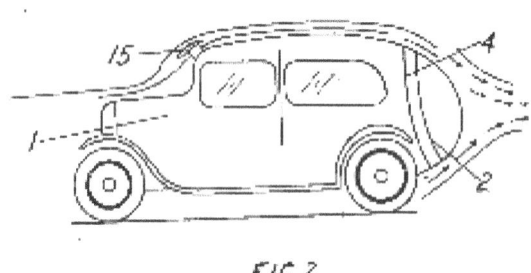

FIG. 2

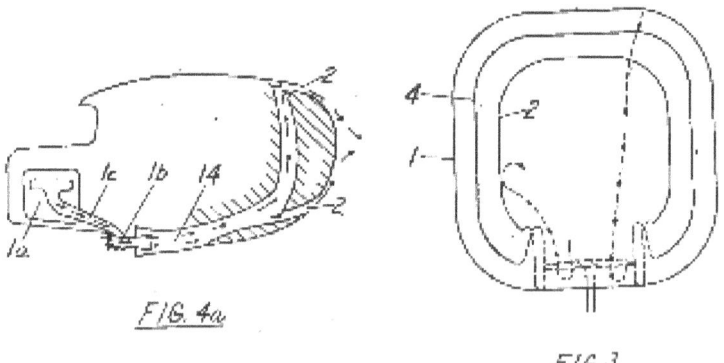

FIG. 4a

FIG. 3

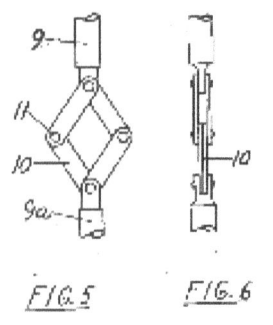

FIG. 5 FIG. 6

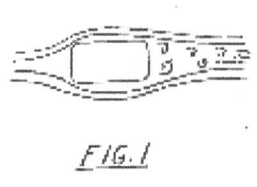

FIG. 1

INVENTOR
Edward A. Stalker

March 6, 1956 R. B. POTTER 2,737,411
INFLATABLE STREAMLINING APPARATUS FOR VEHICLE BODIES
Filed Aug. 21, 1952

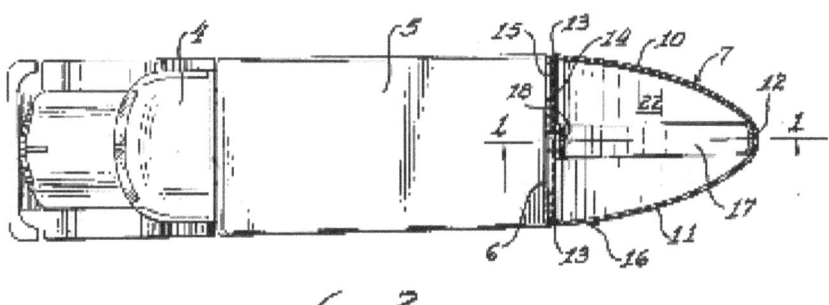

Fig. 2

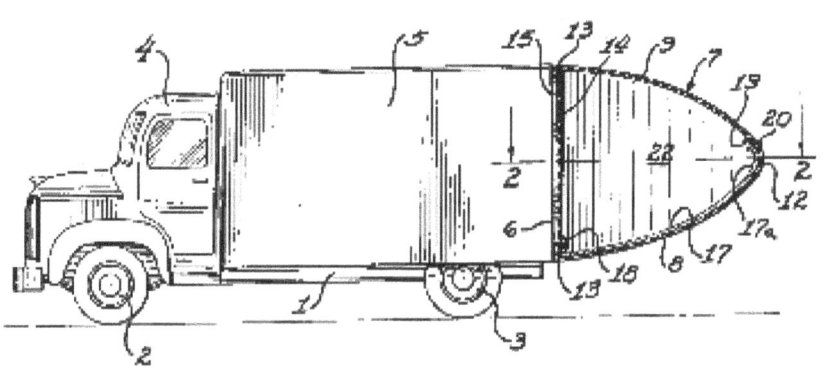

Fig. 1

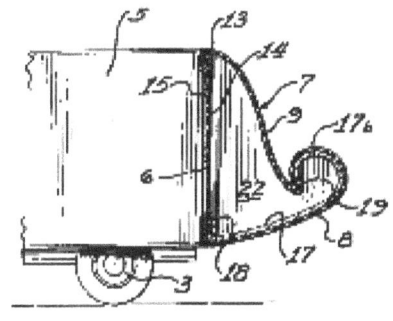

Fig. 3

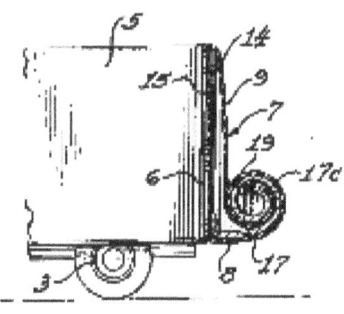

Fig. 4

INVENTOR.
RALPH B. POTTER.

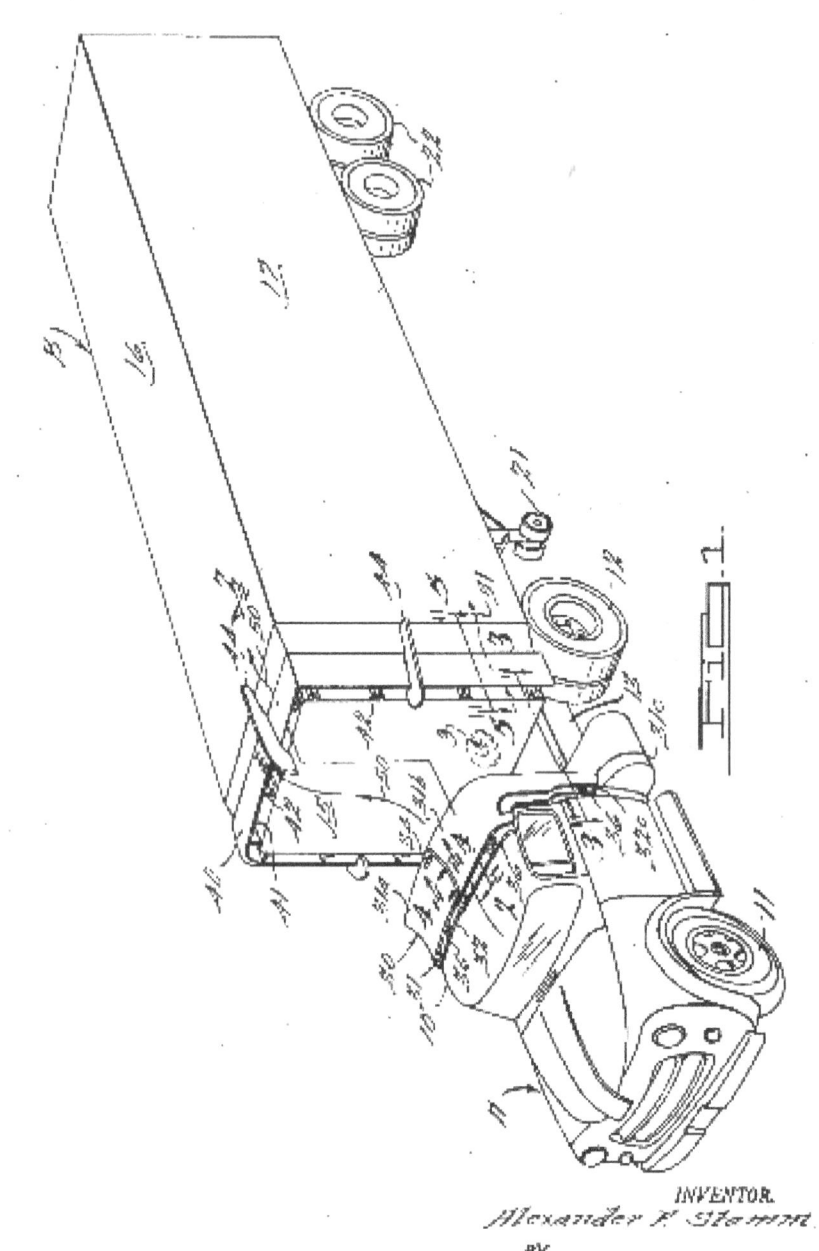

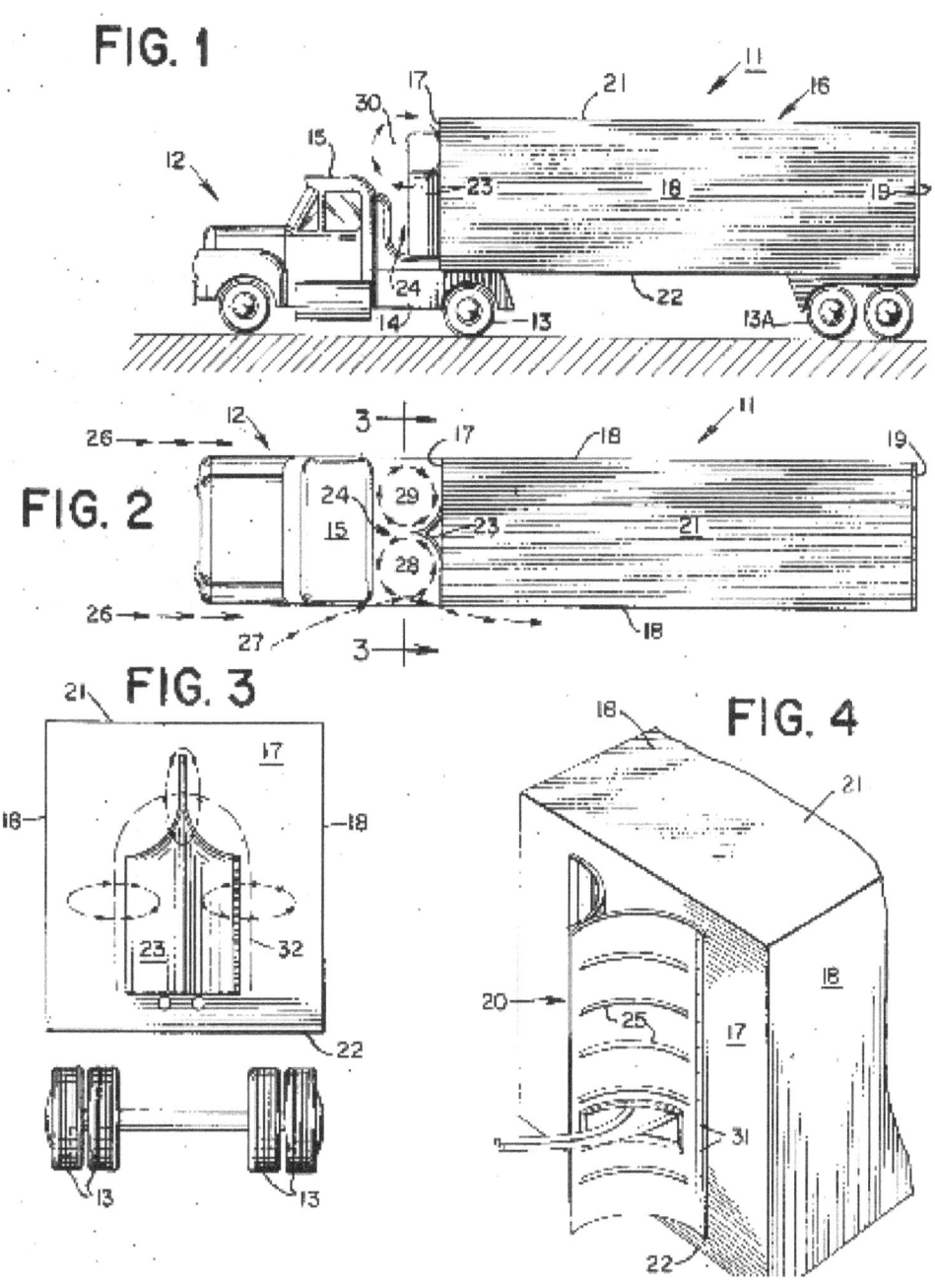

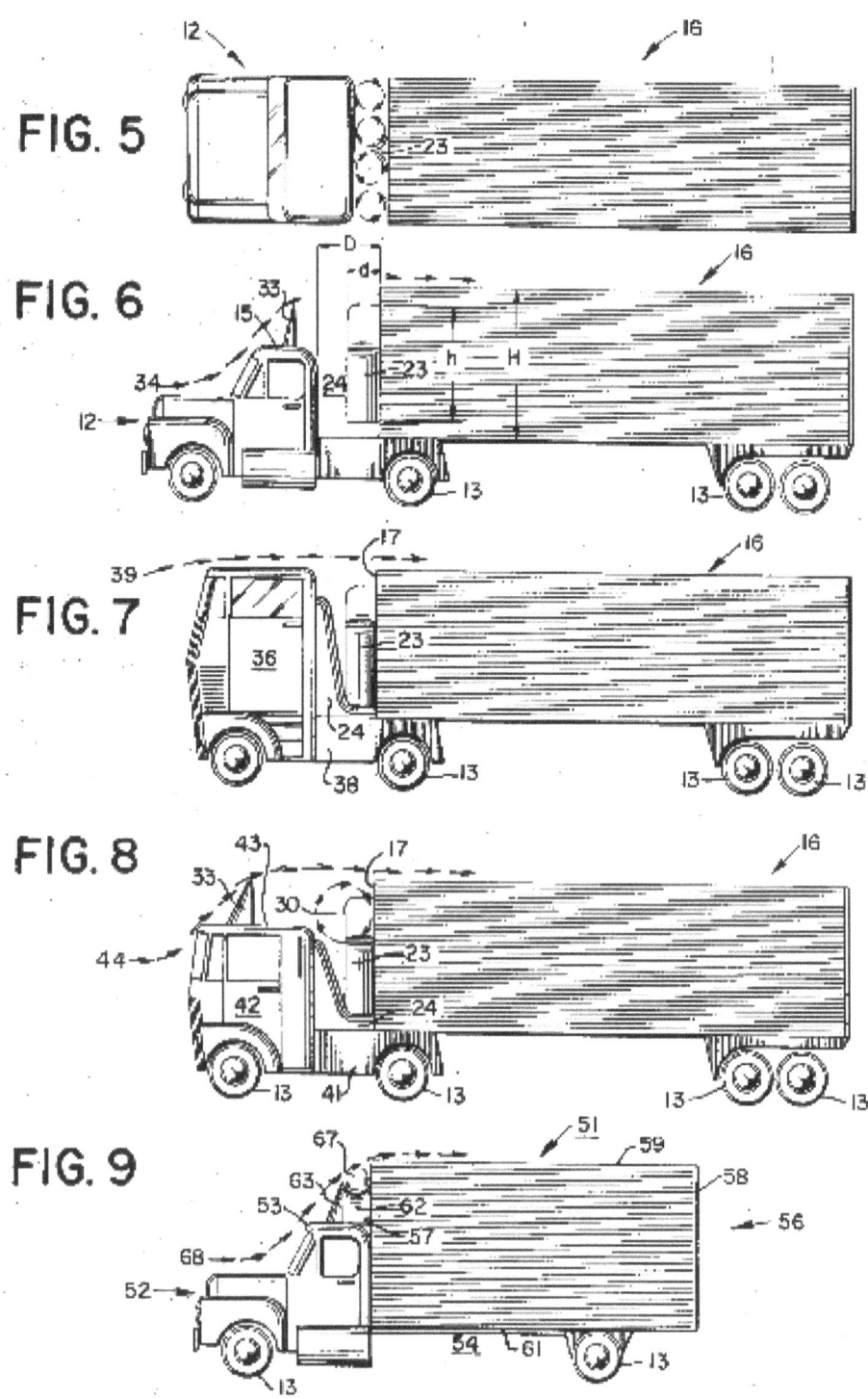

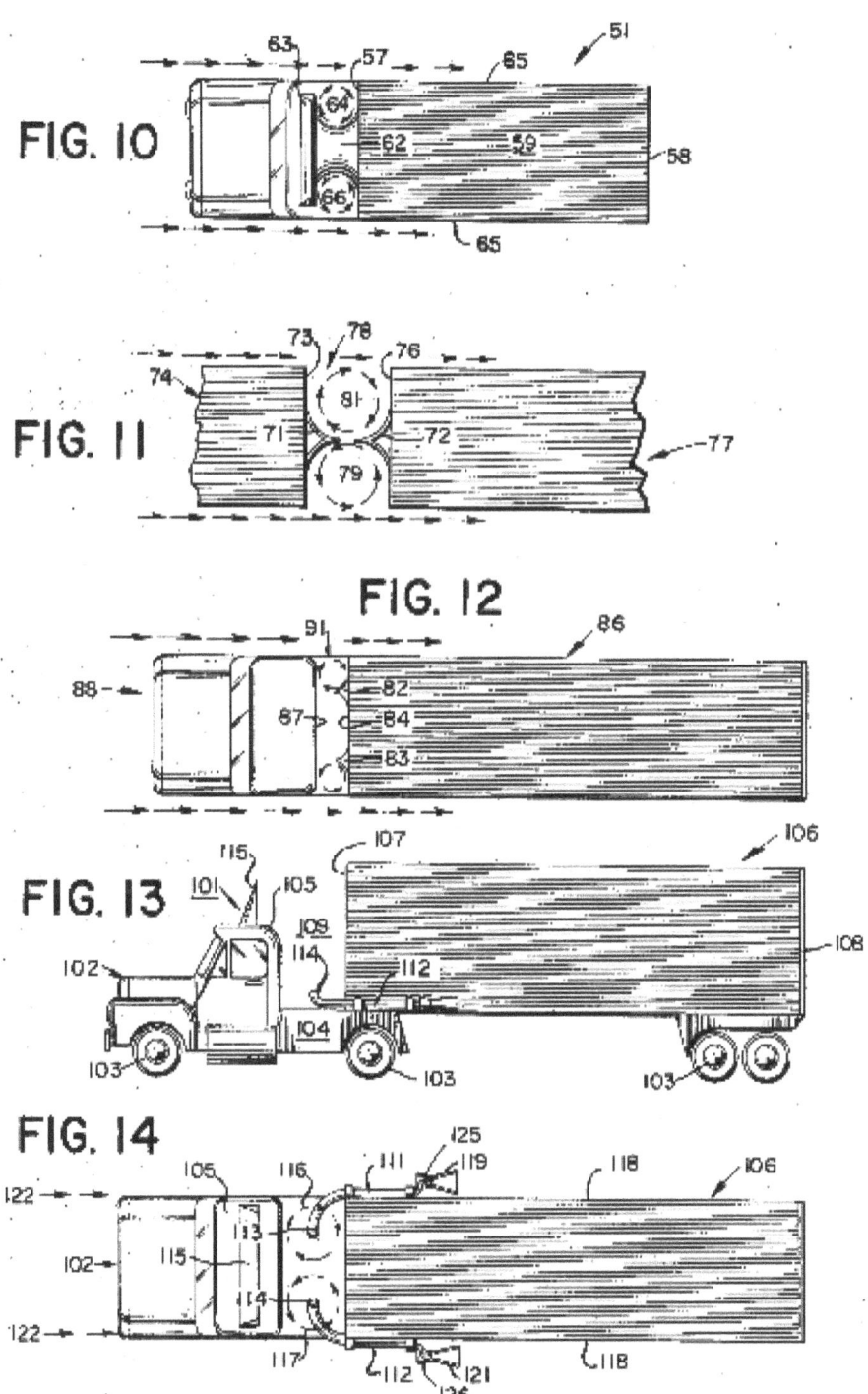

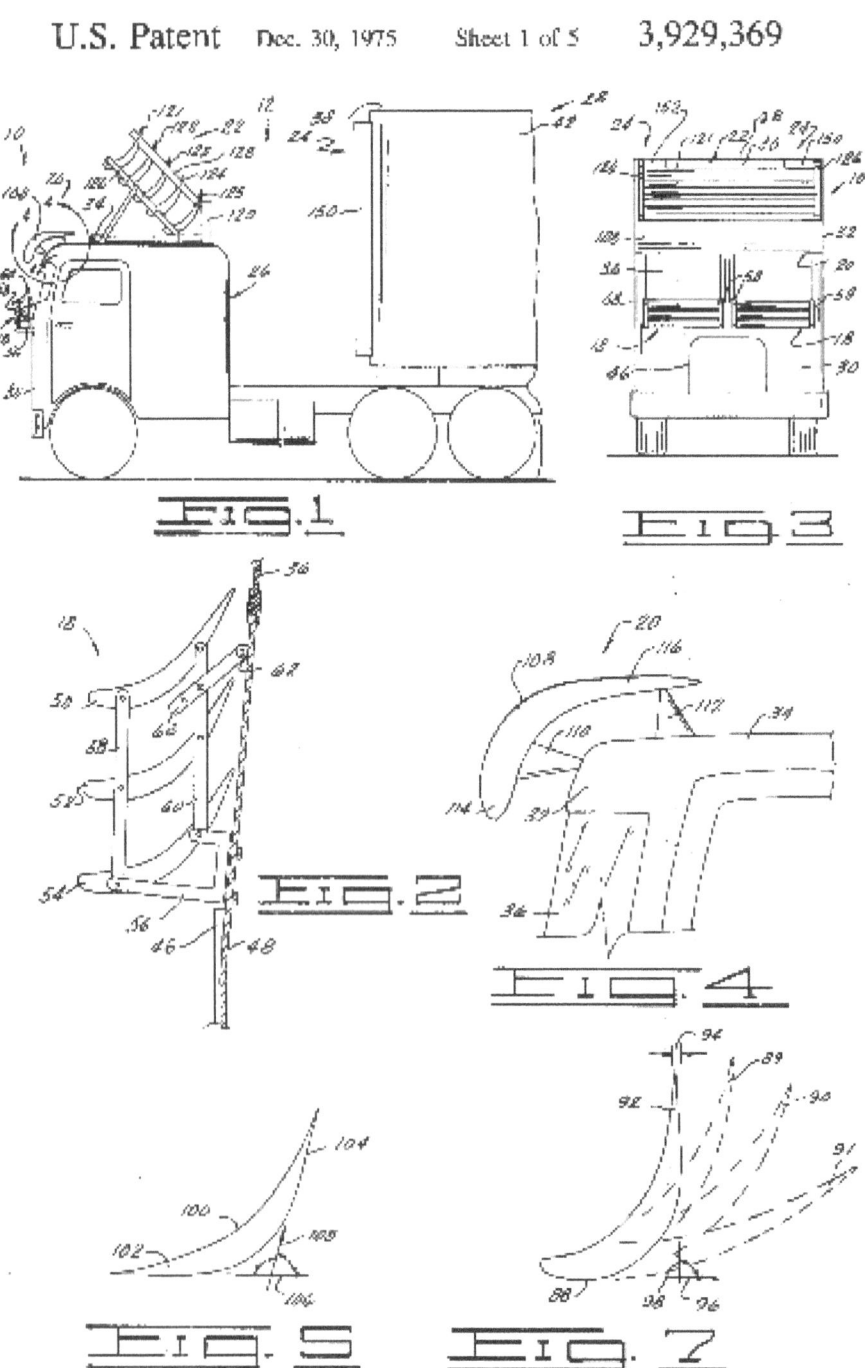

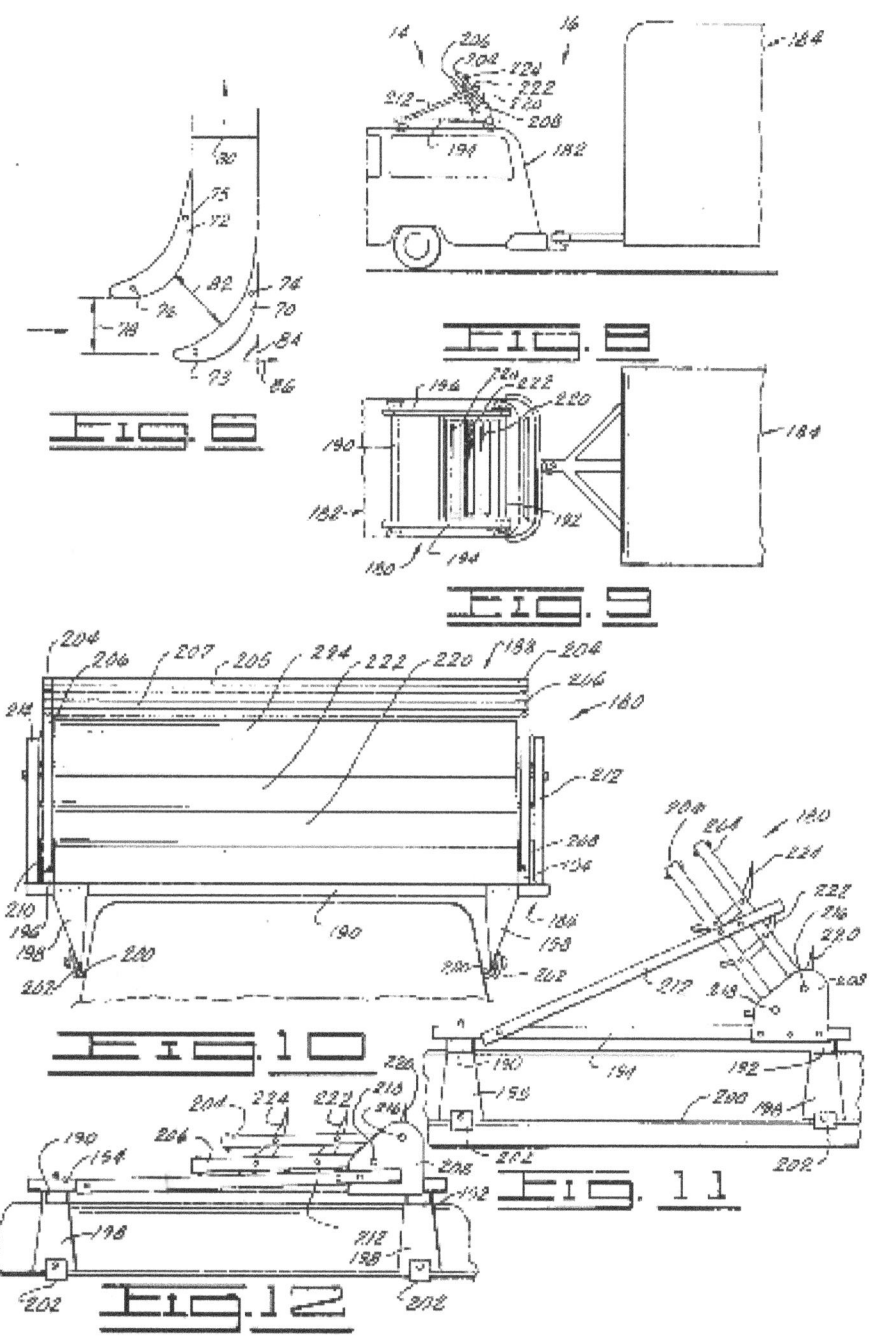

Appendices

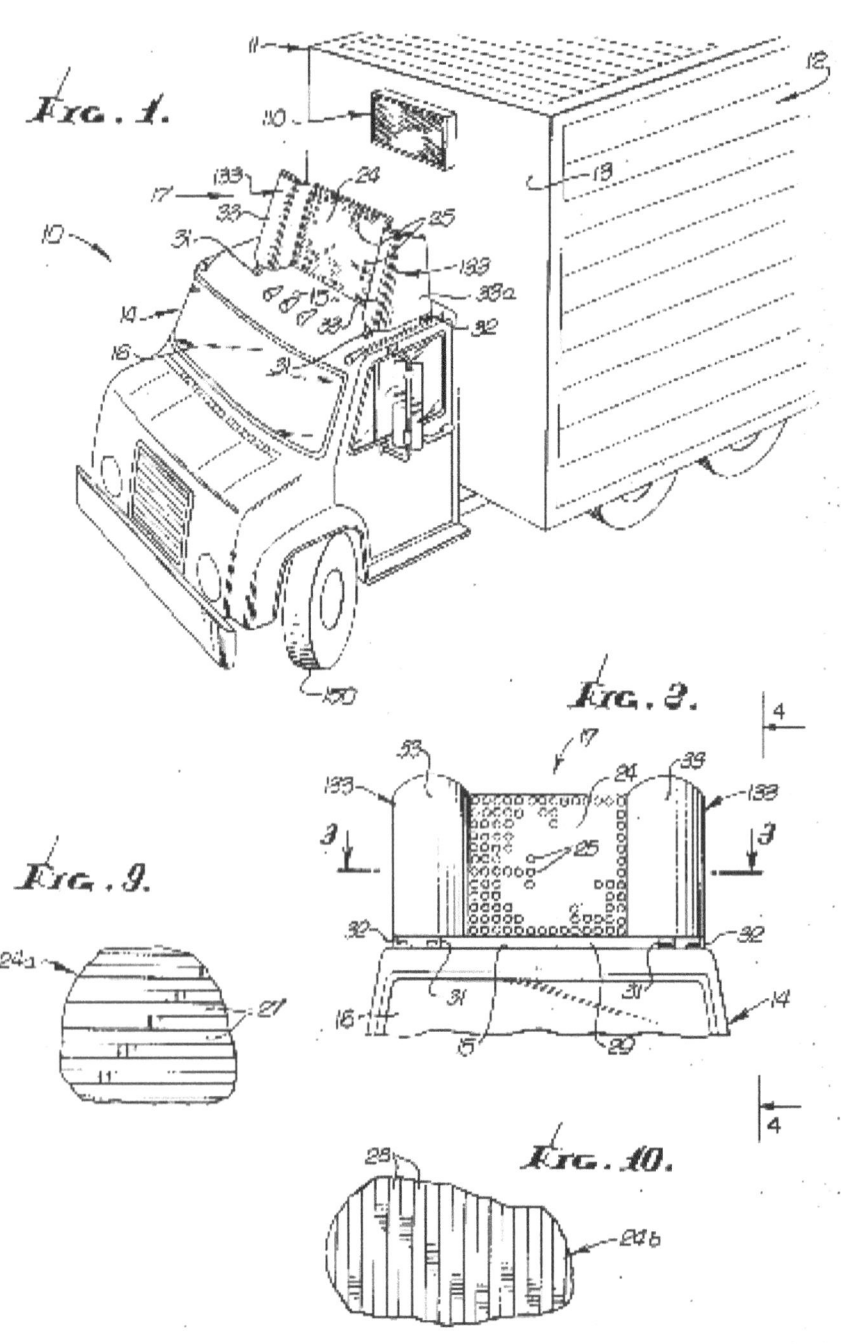

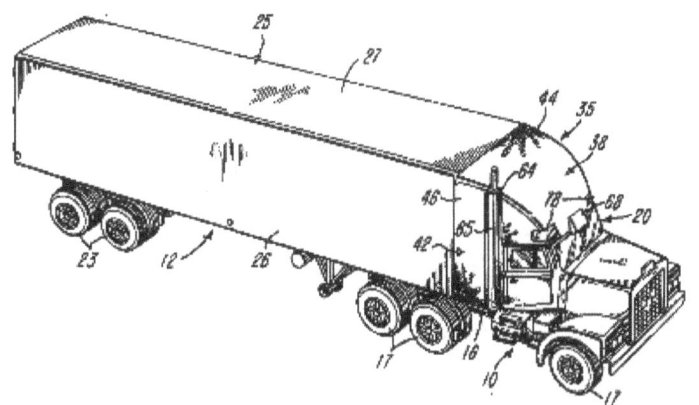

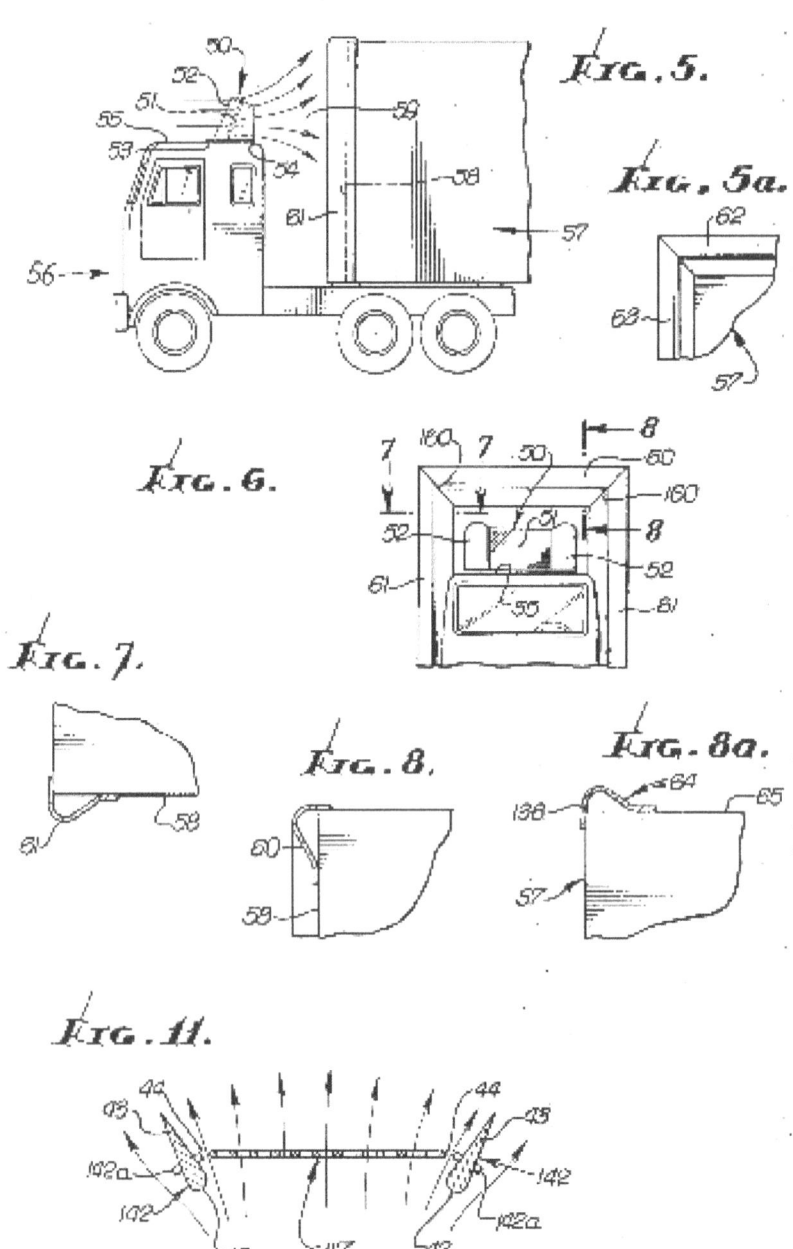

95

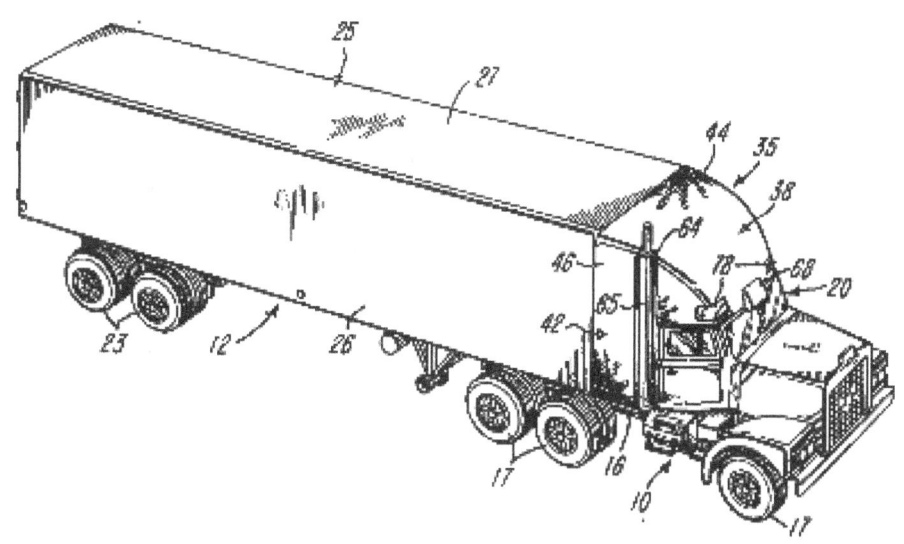

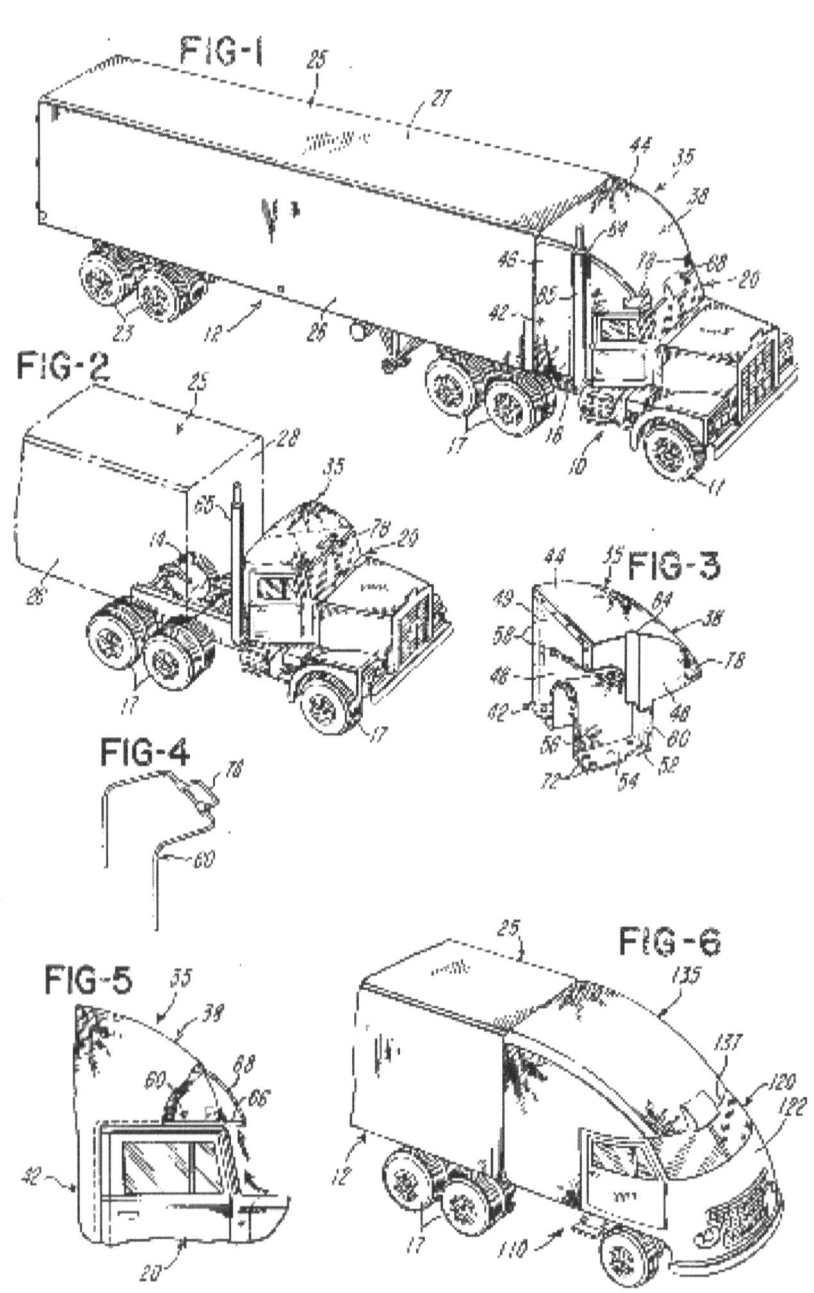

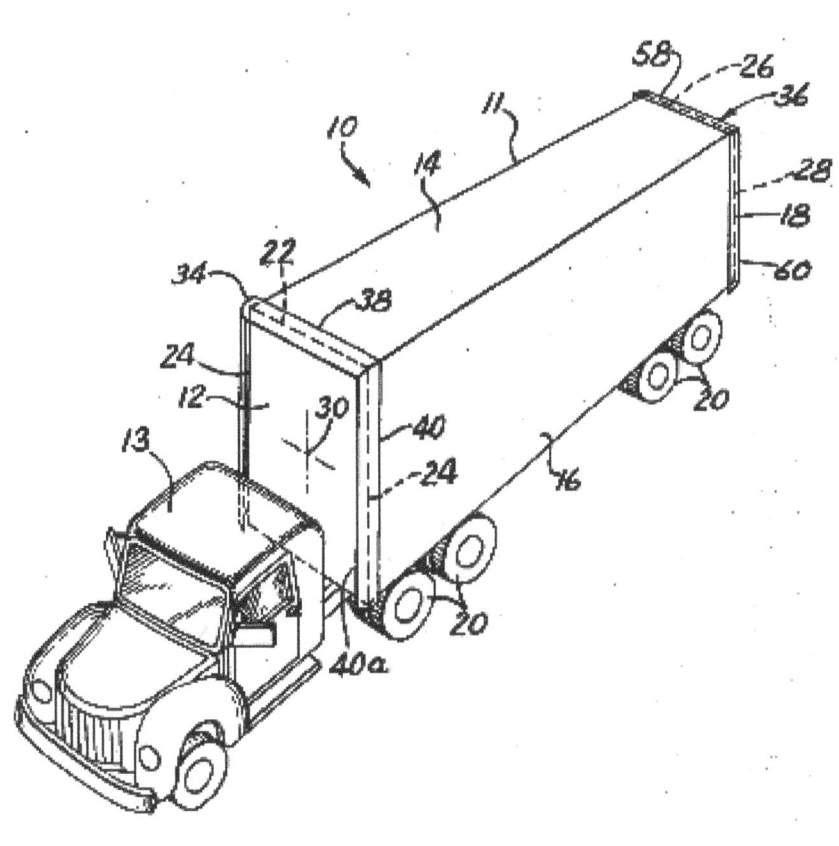

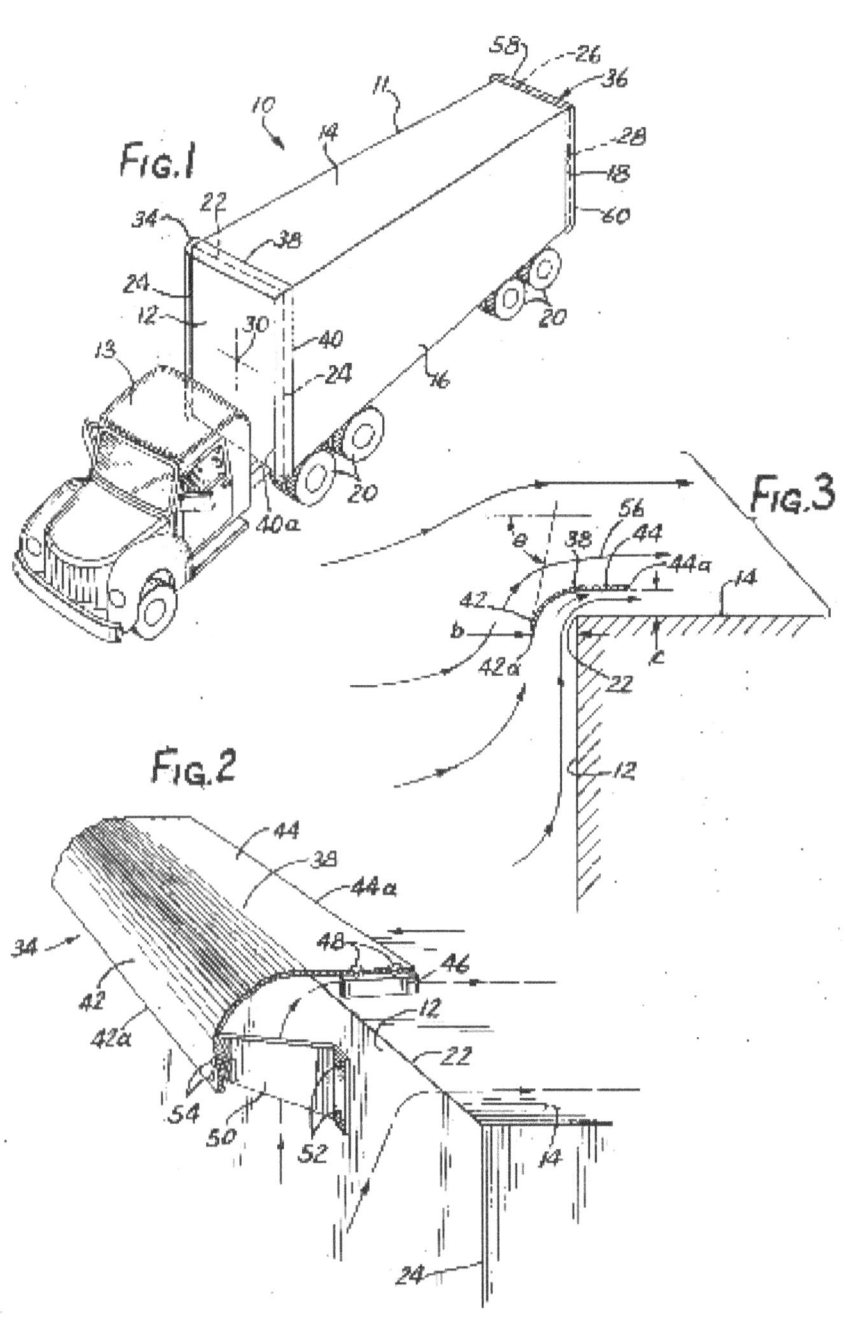

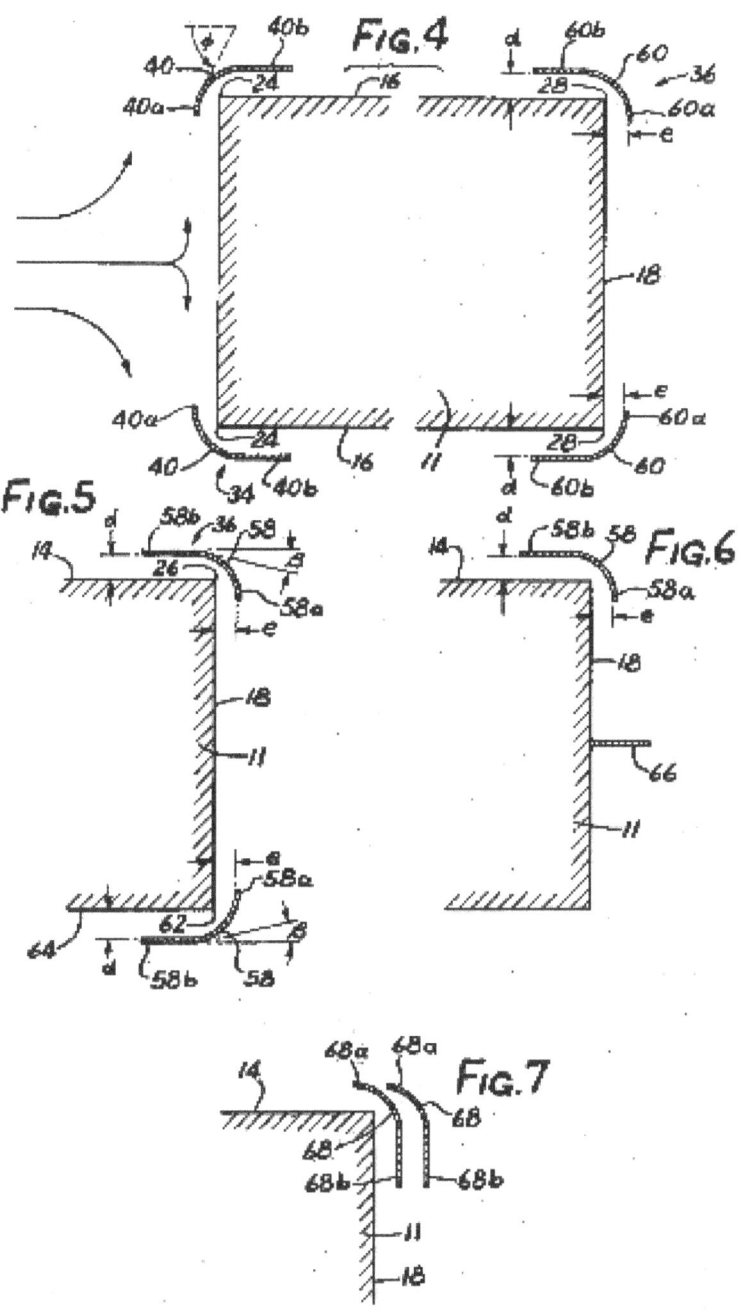

Appendices

U.S. Patent   Feb. 8, 1977   4,006,931

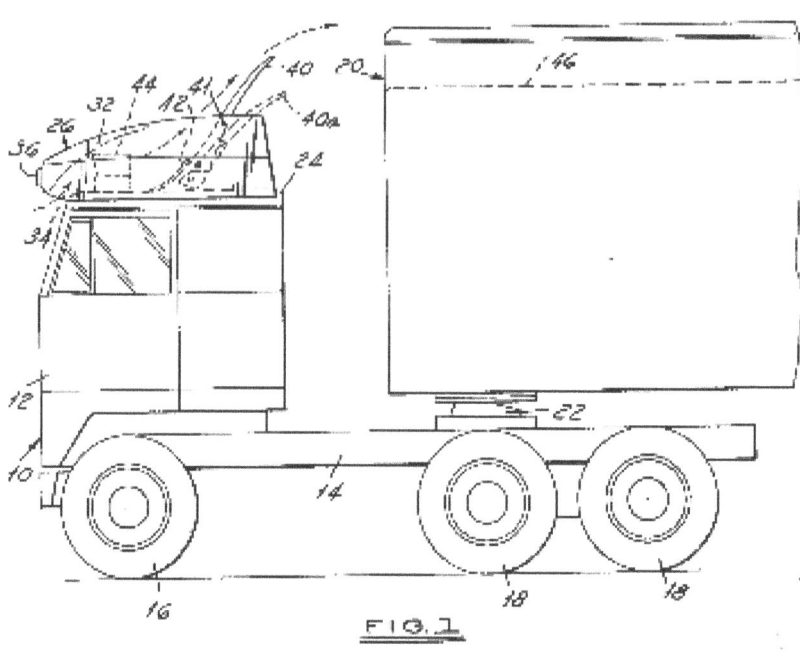

FIG. 1

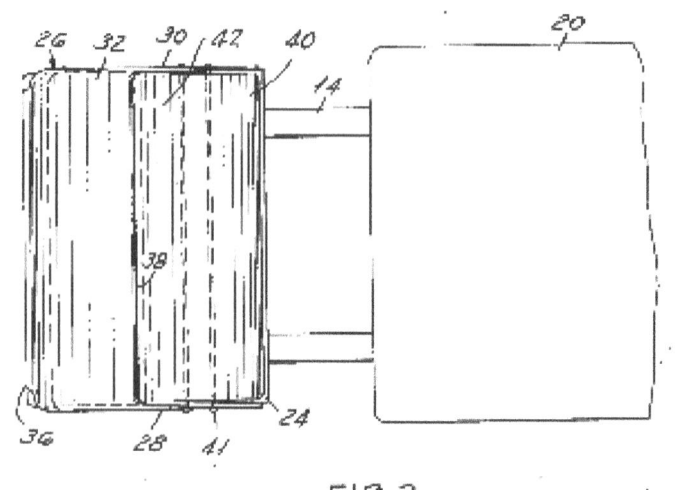

FIG. 2

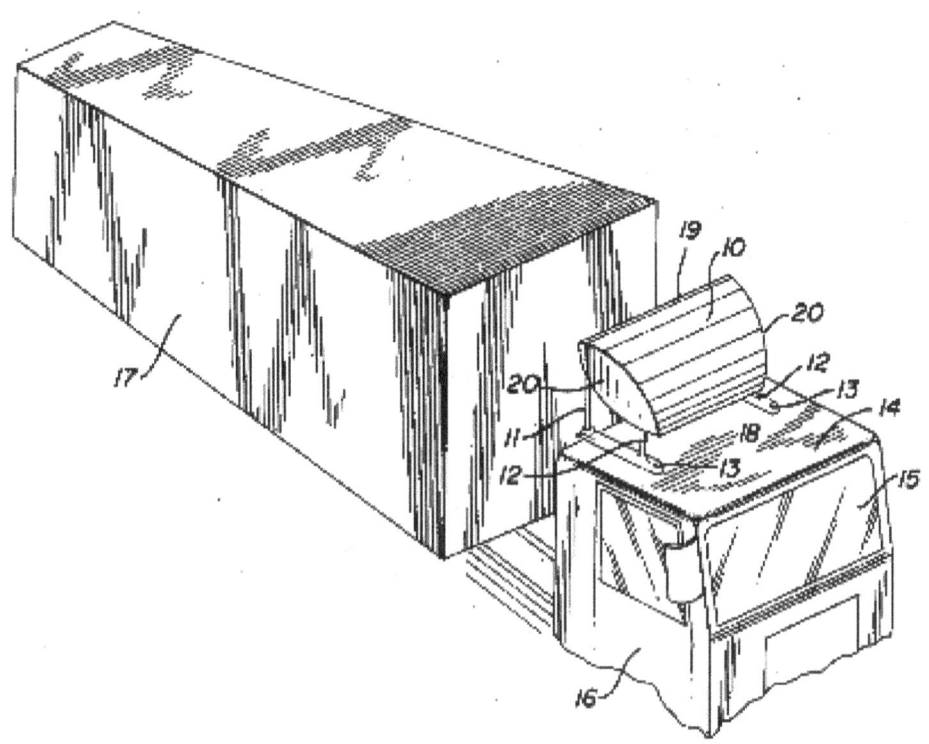

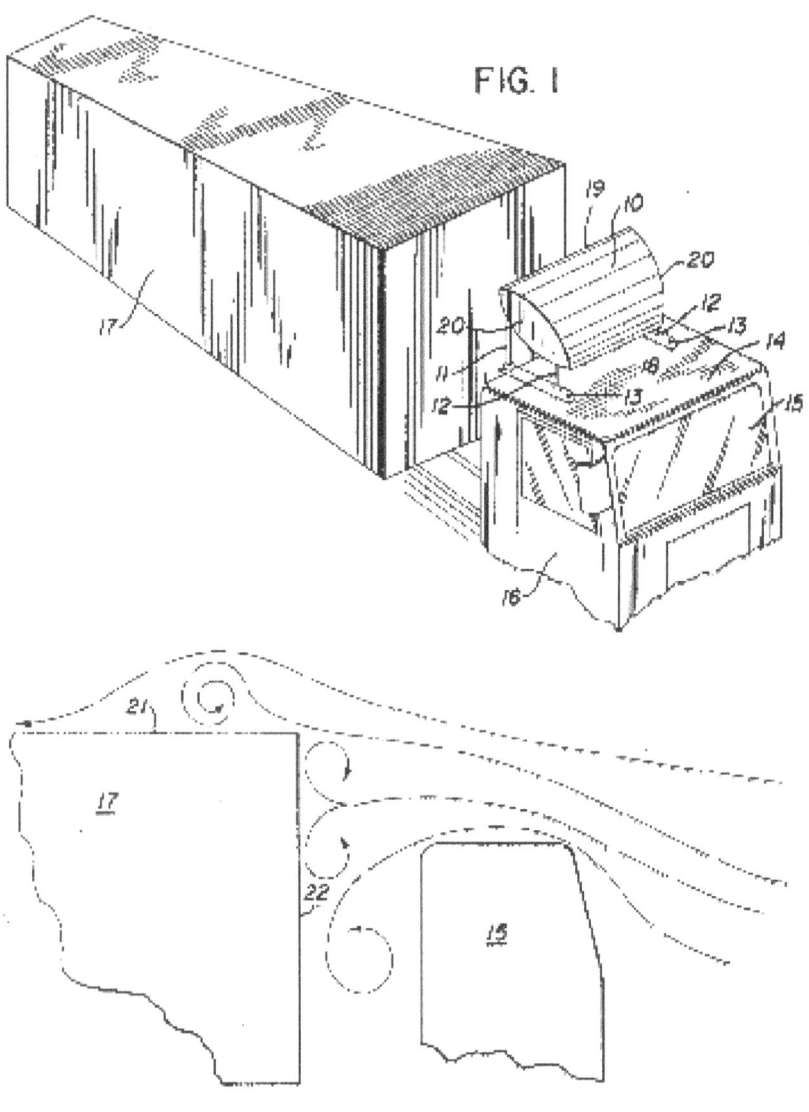

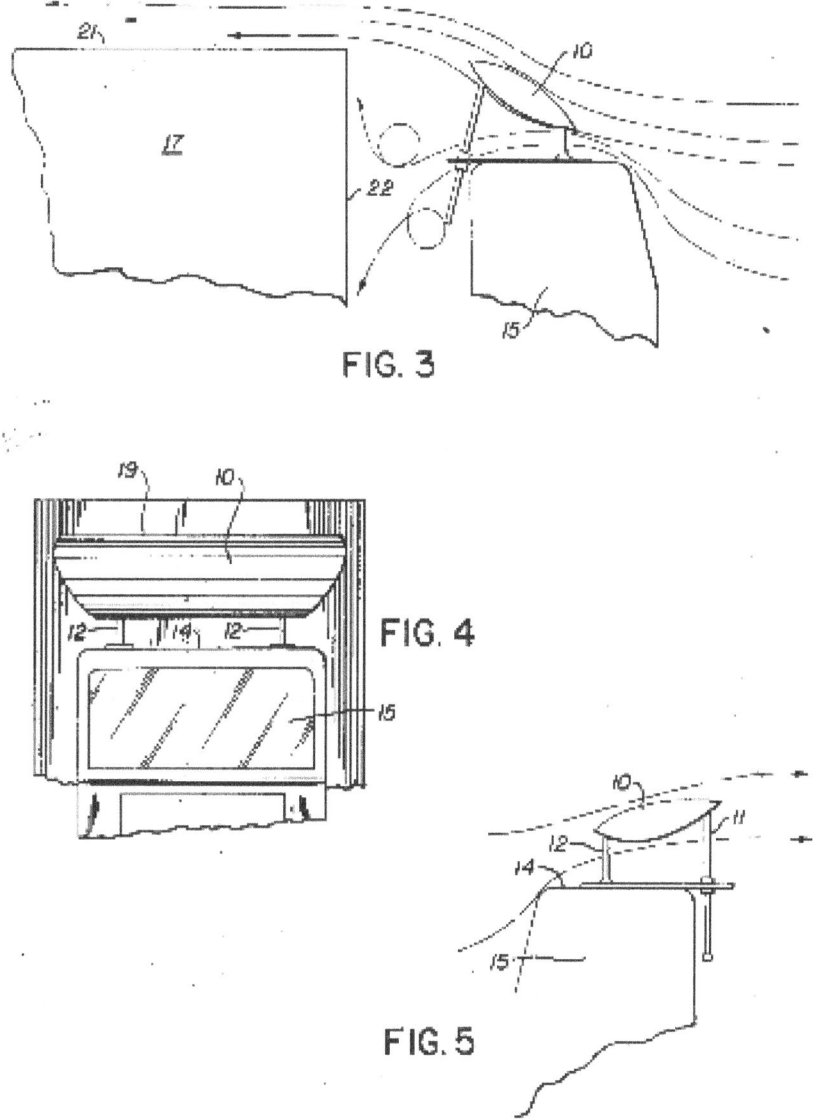

FIG. 3

FIG. 4

FIG. 5

104 Appendices

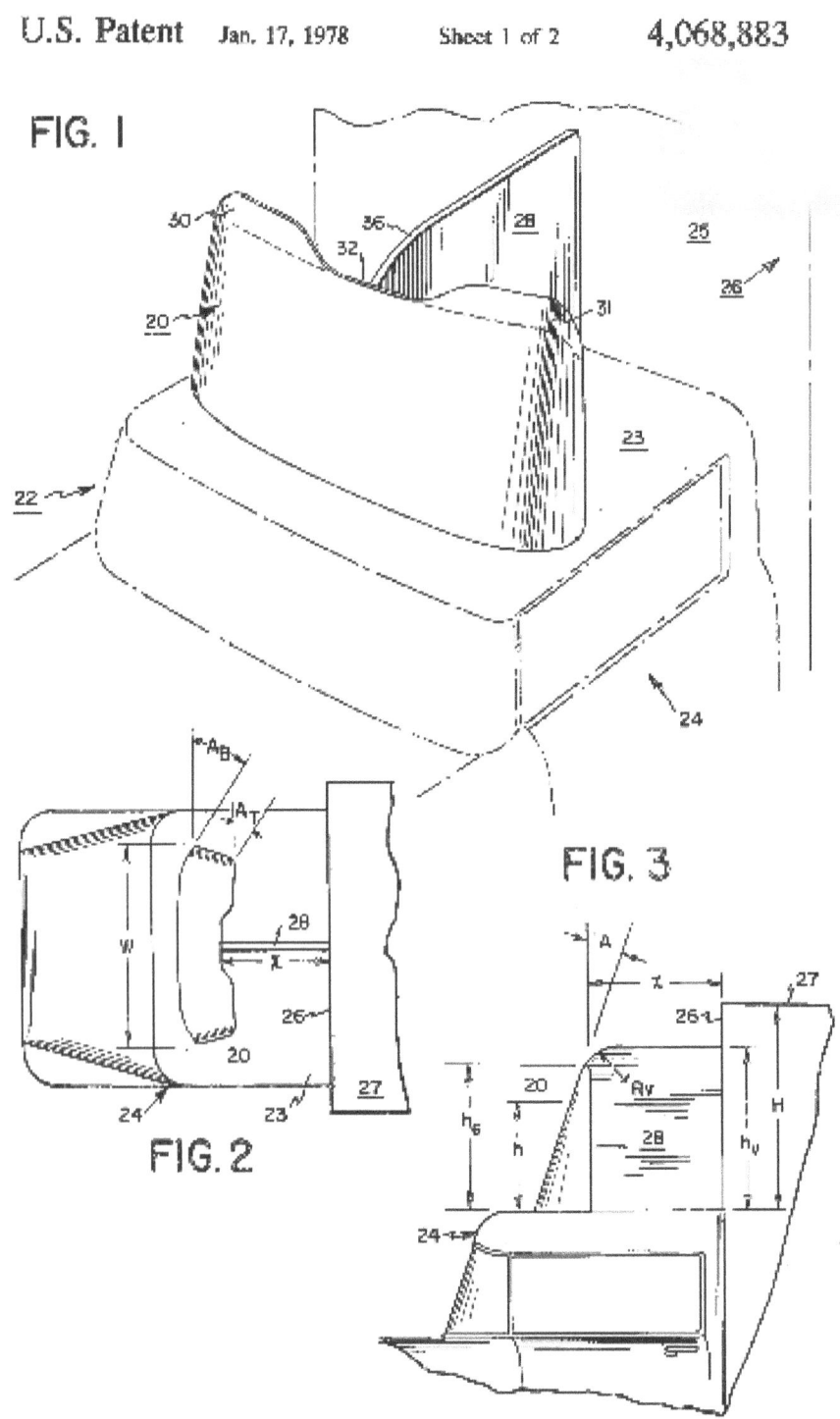

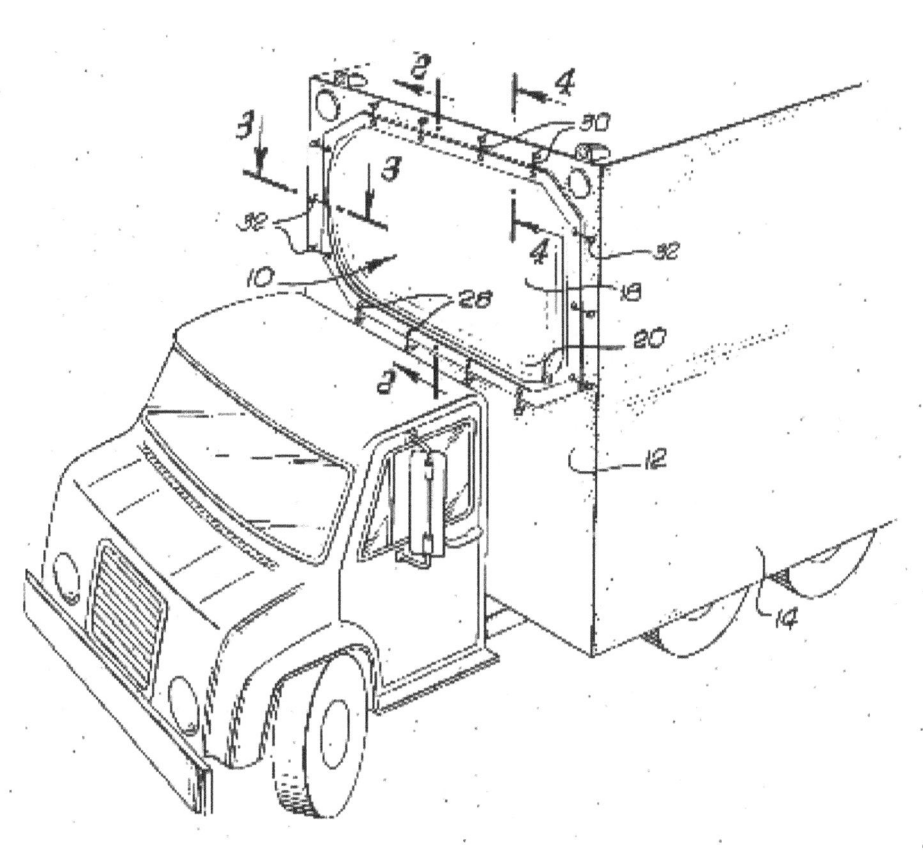

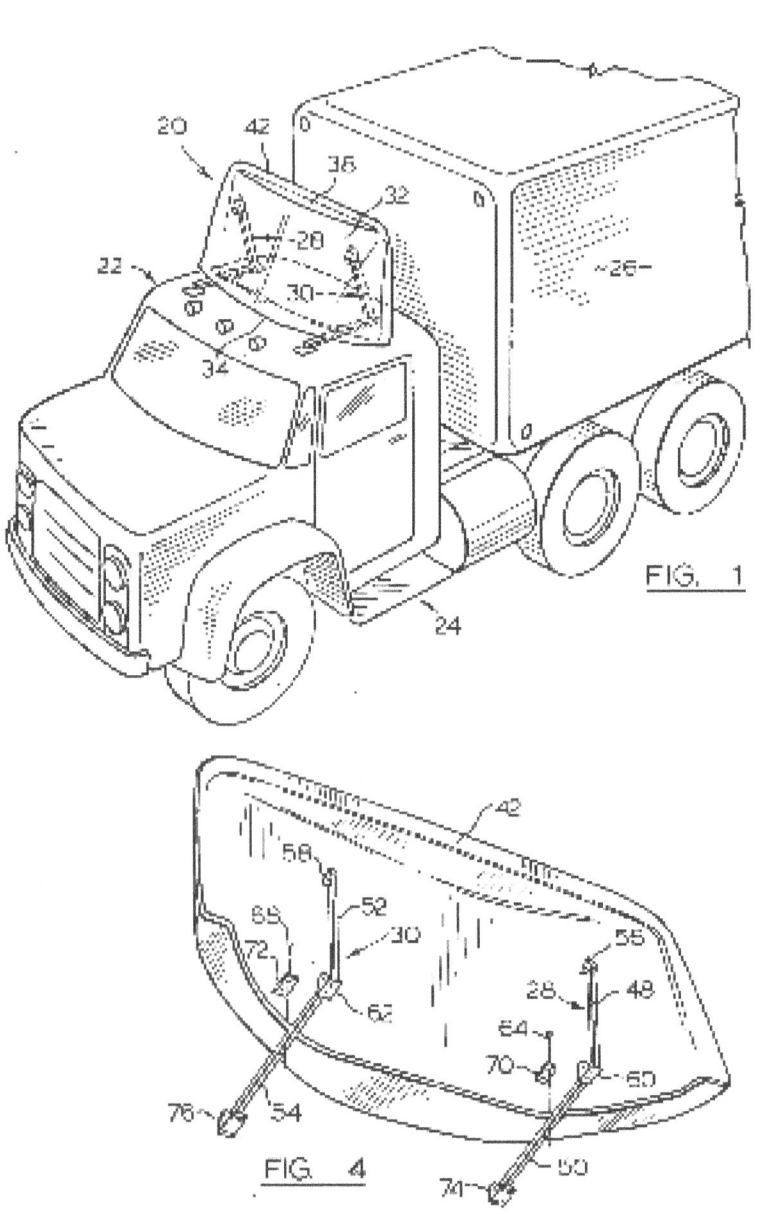

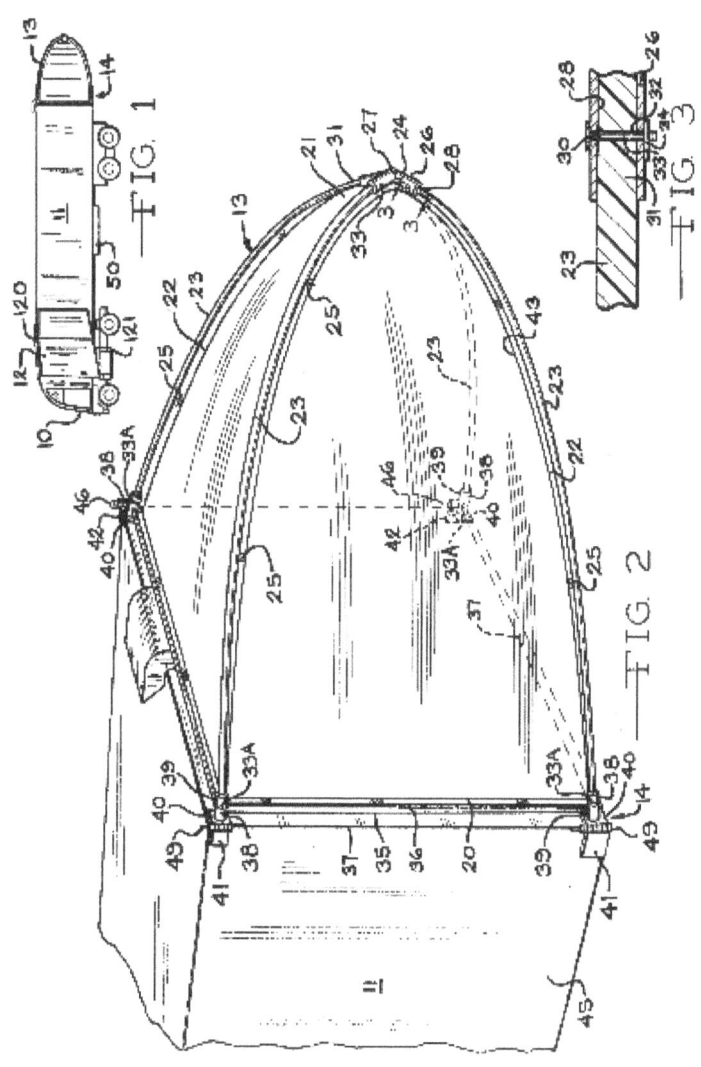

108 Appendices

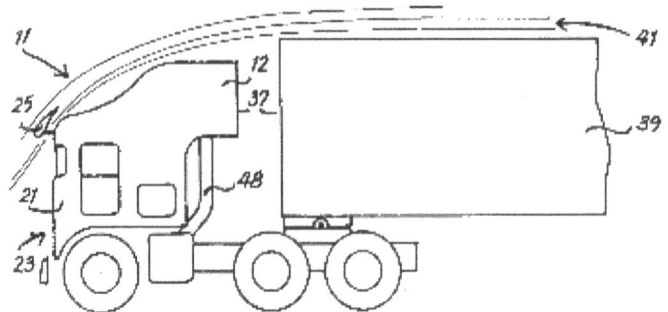

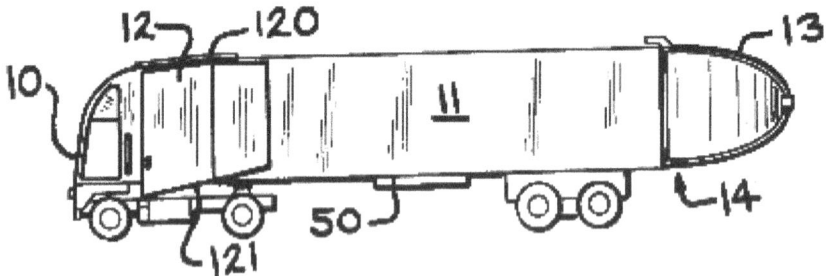

U.S. Patent   Apr. 22, 1980   Sheet 1 of 2   4,199,185

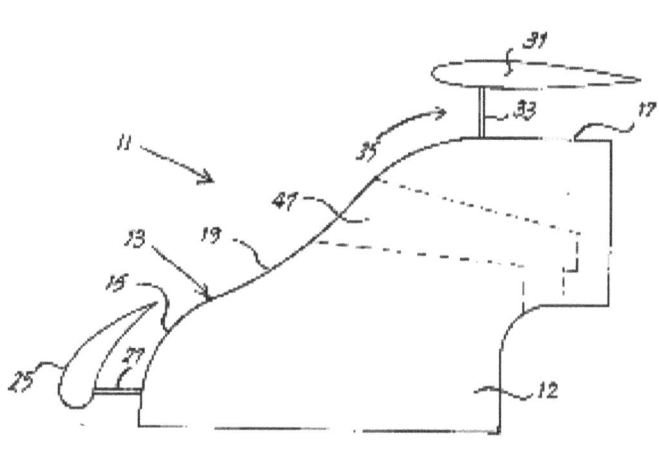

Fig. 1

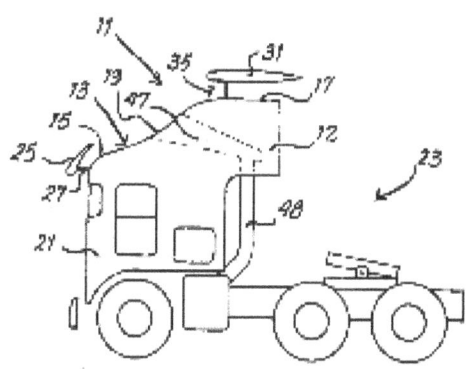

Fig. 2

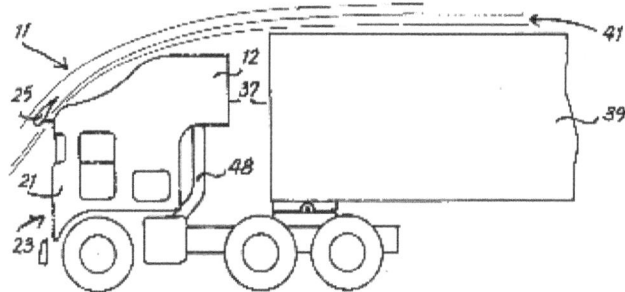

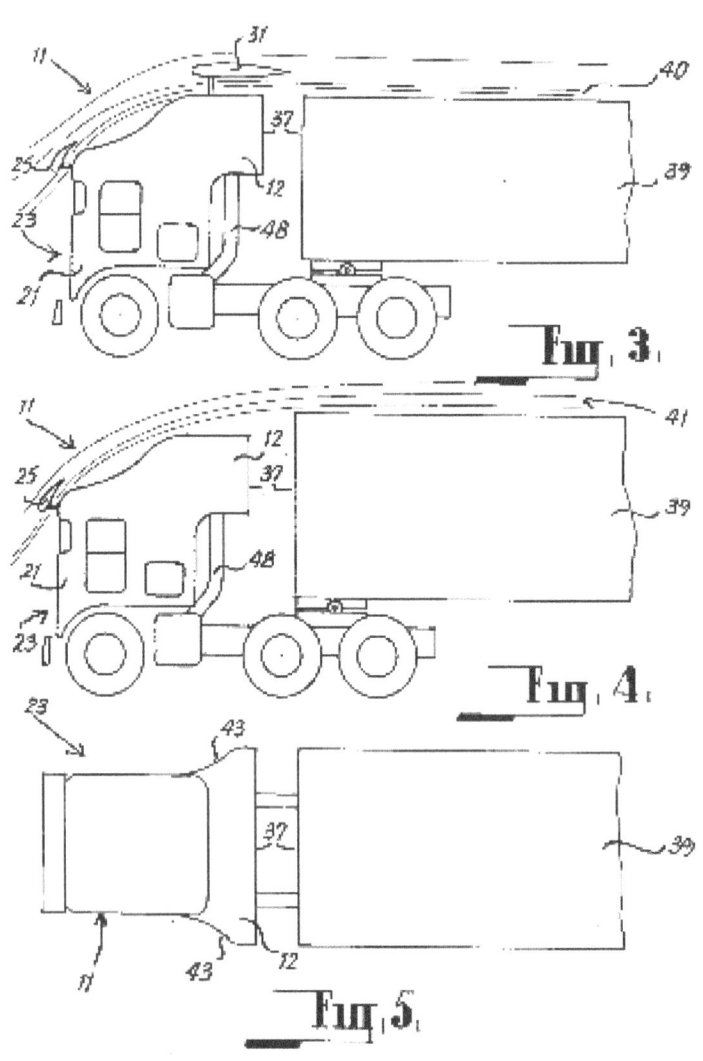

Fig. 3.

Fig. 4.

Fig. 5.

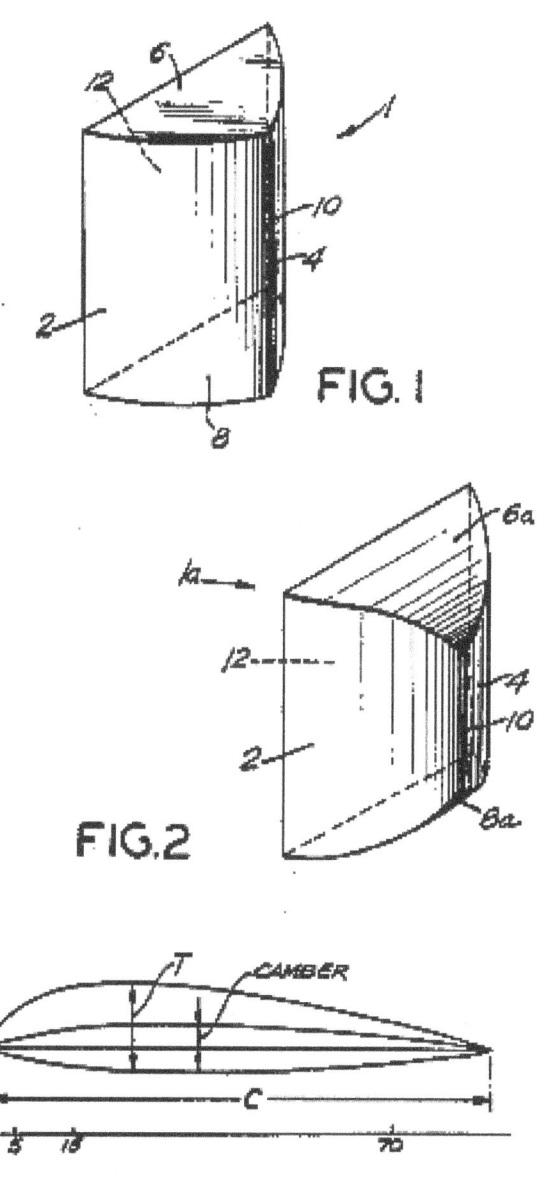

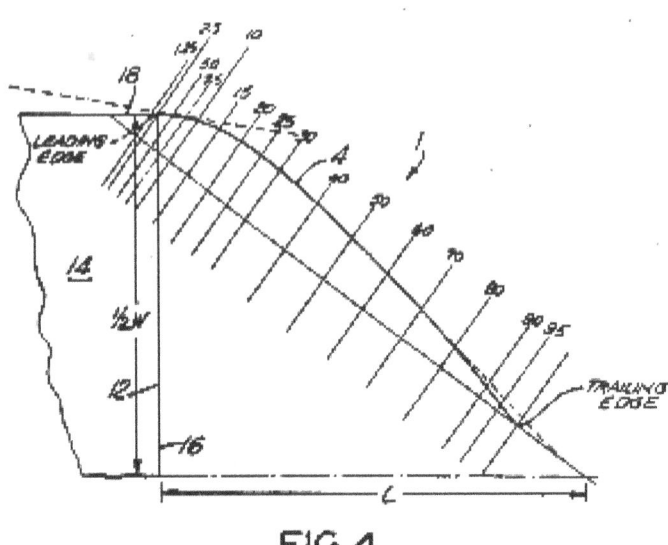

FIG. 4

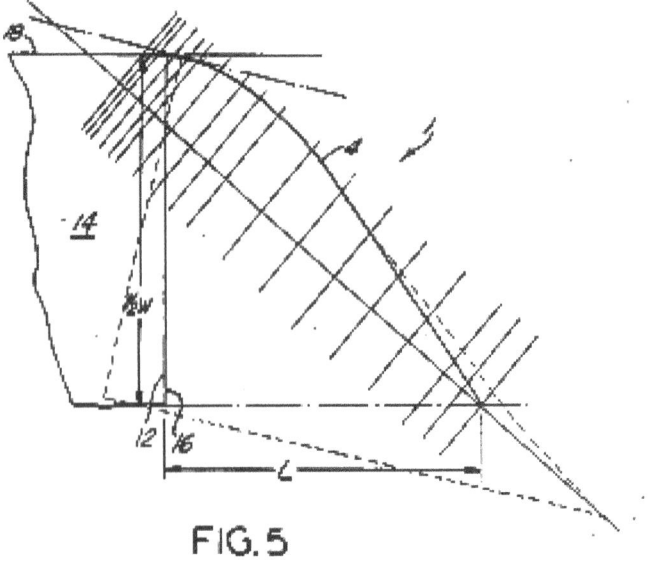

FIG. 5

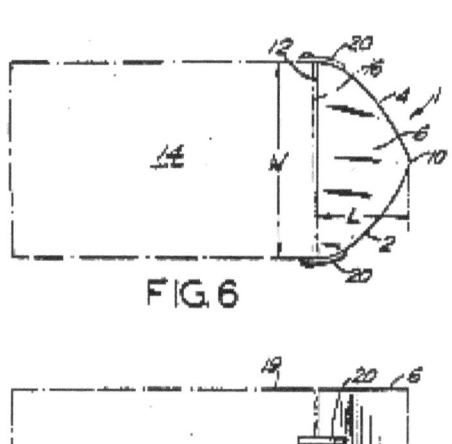

FIG. 6

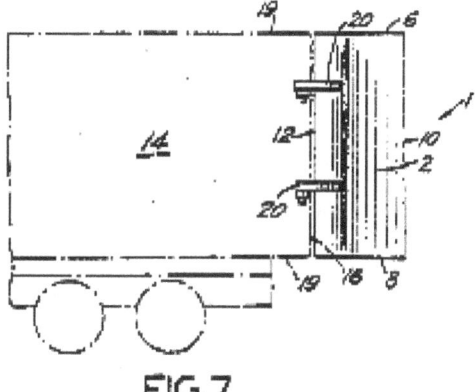

FIG. 7

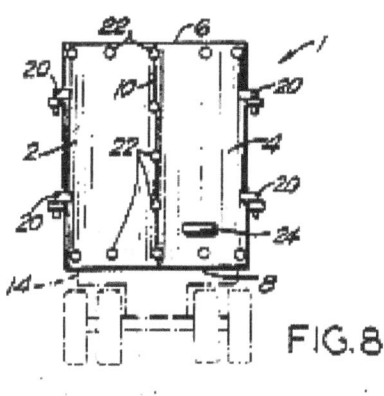

FIG. 8

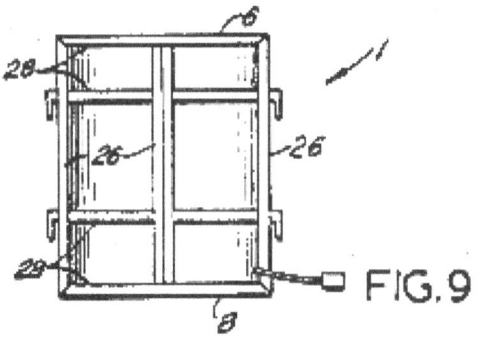

FIG. 9

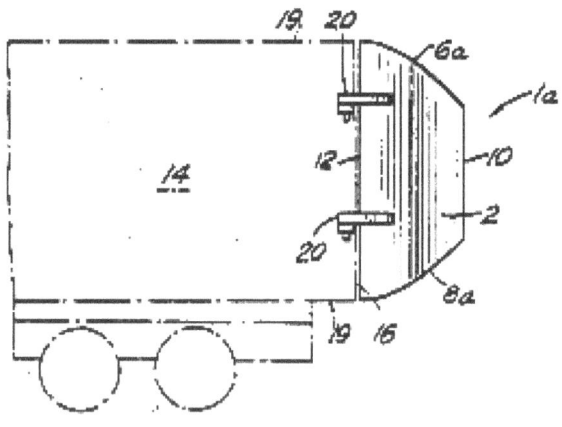

FIG. 10

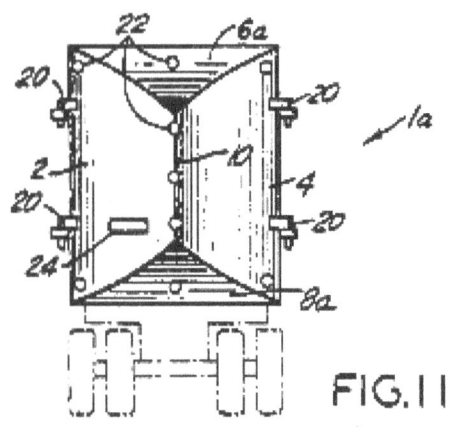

FIG. 11

U.S. Patent Oct. 27, 1987 Sheet 1 of 9 4,702,509

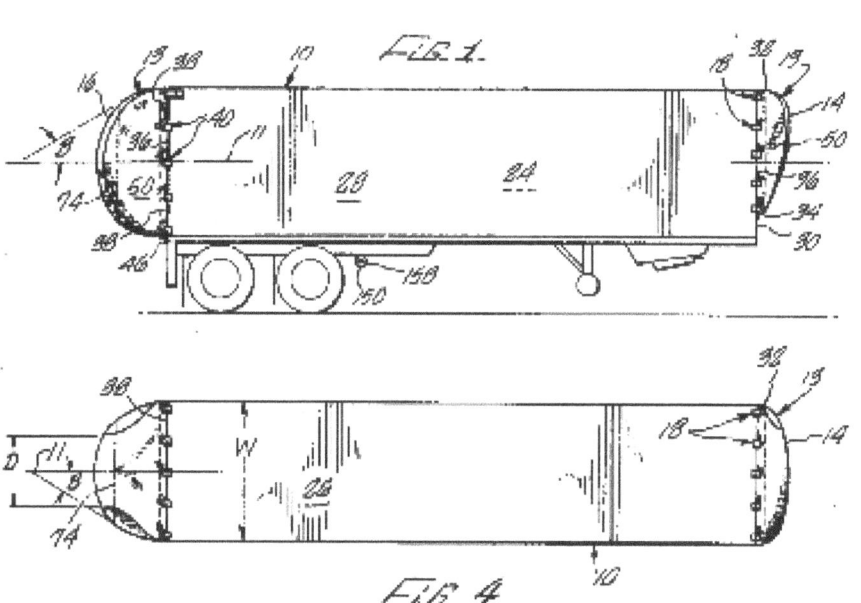

FIG. 1.

FIG. 4.

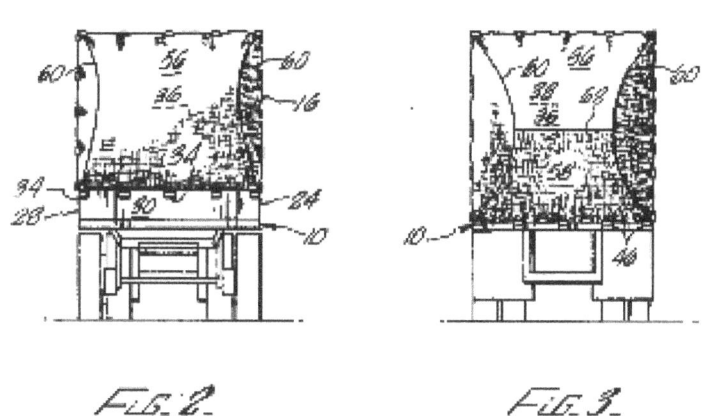

FIG. 2.

FIG. 3.

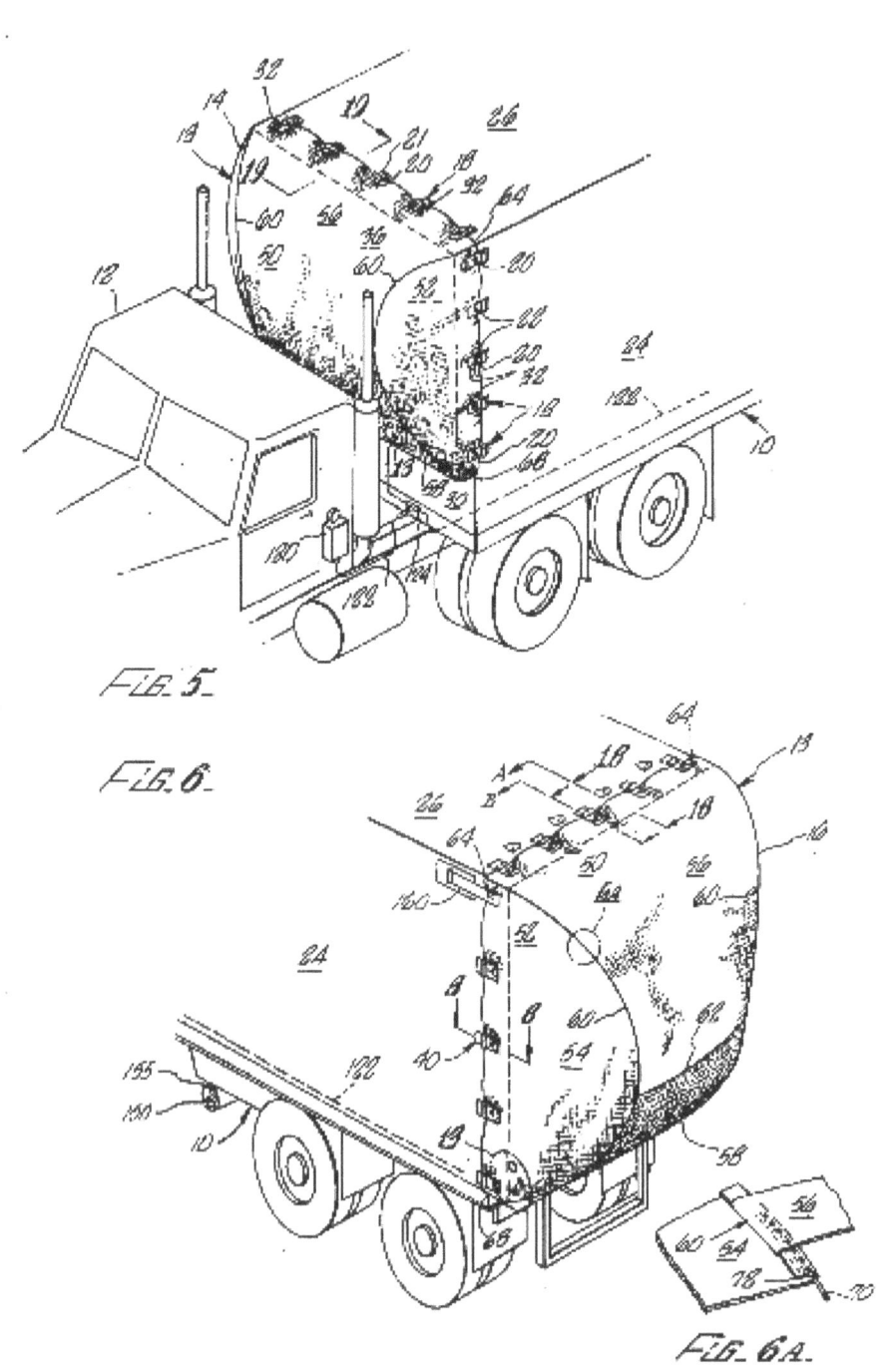

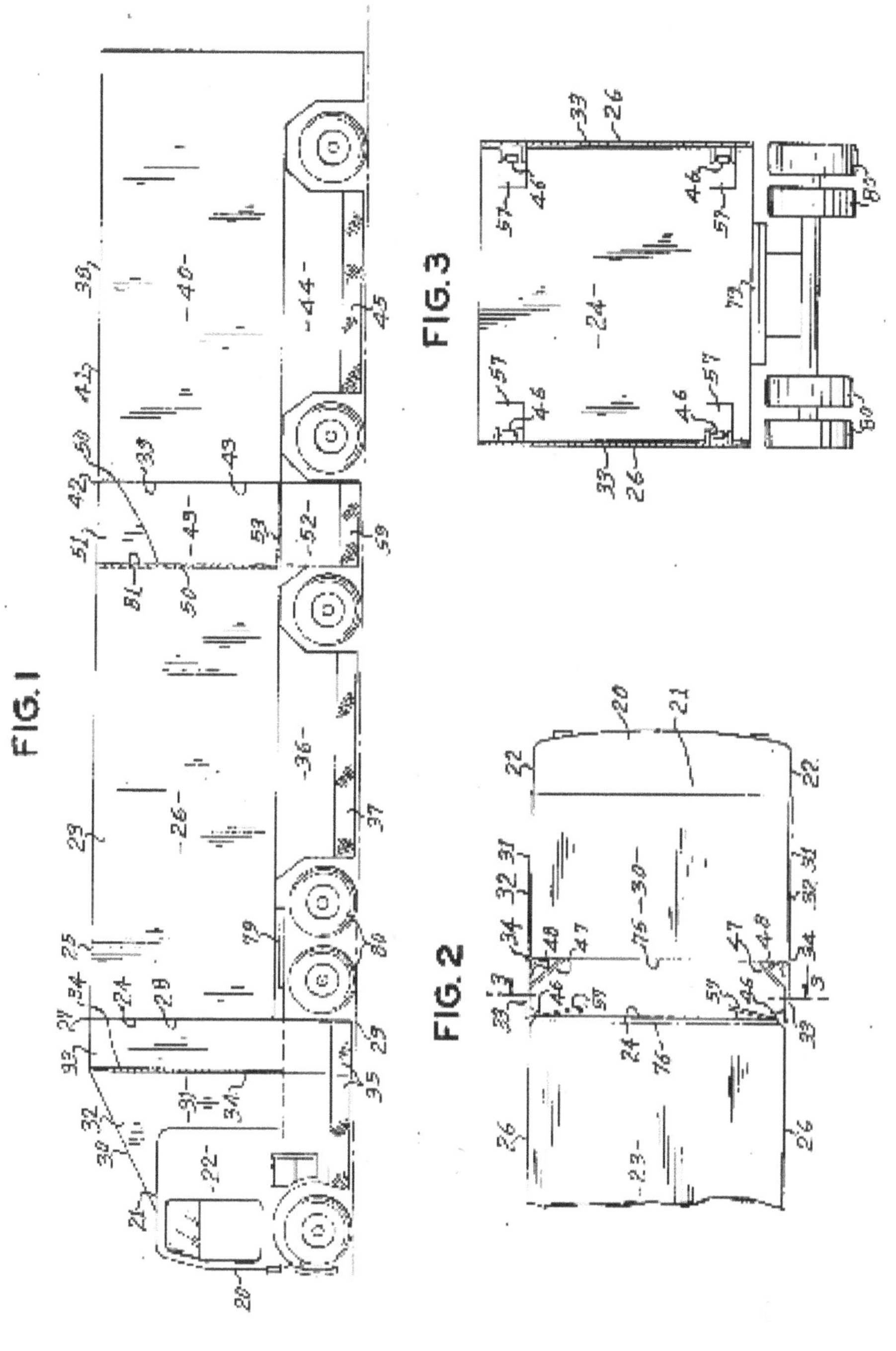

FIG. 1

FIG. 2

FIG. 3

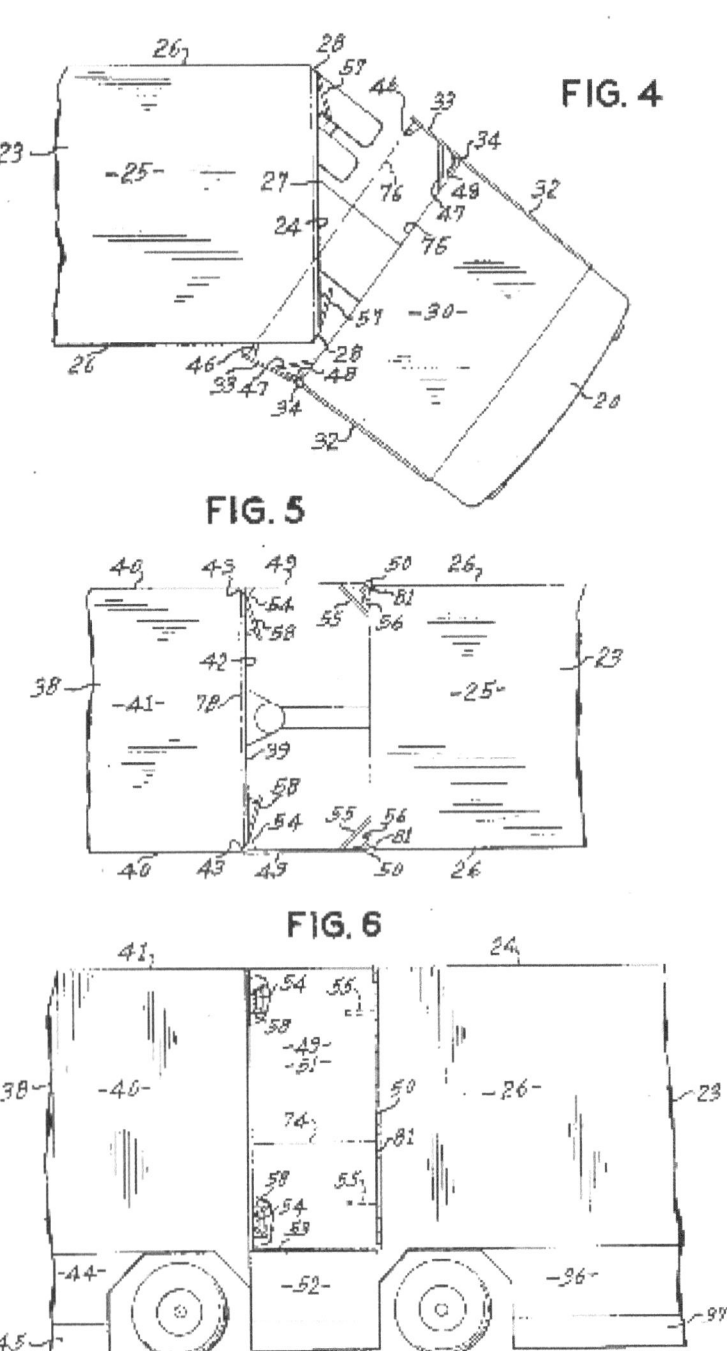

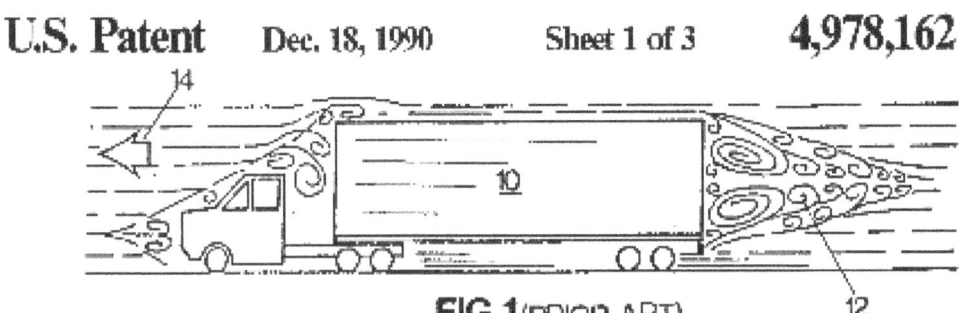

FIG.1 (PRIOR ART)

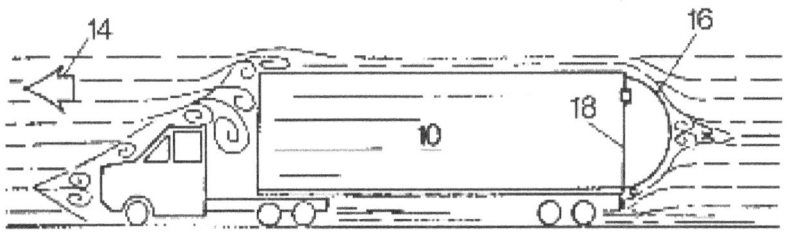

FIG.2

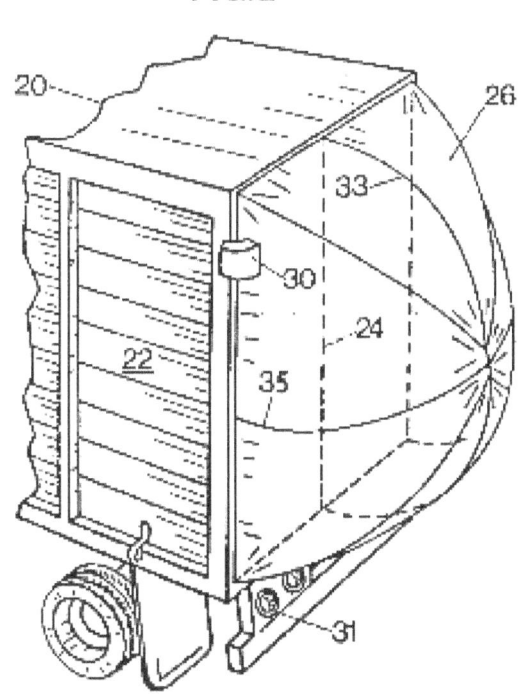

FIG.3

Appendices

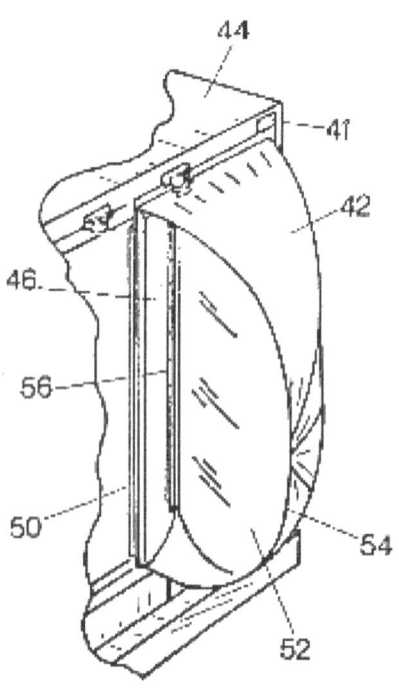

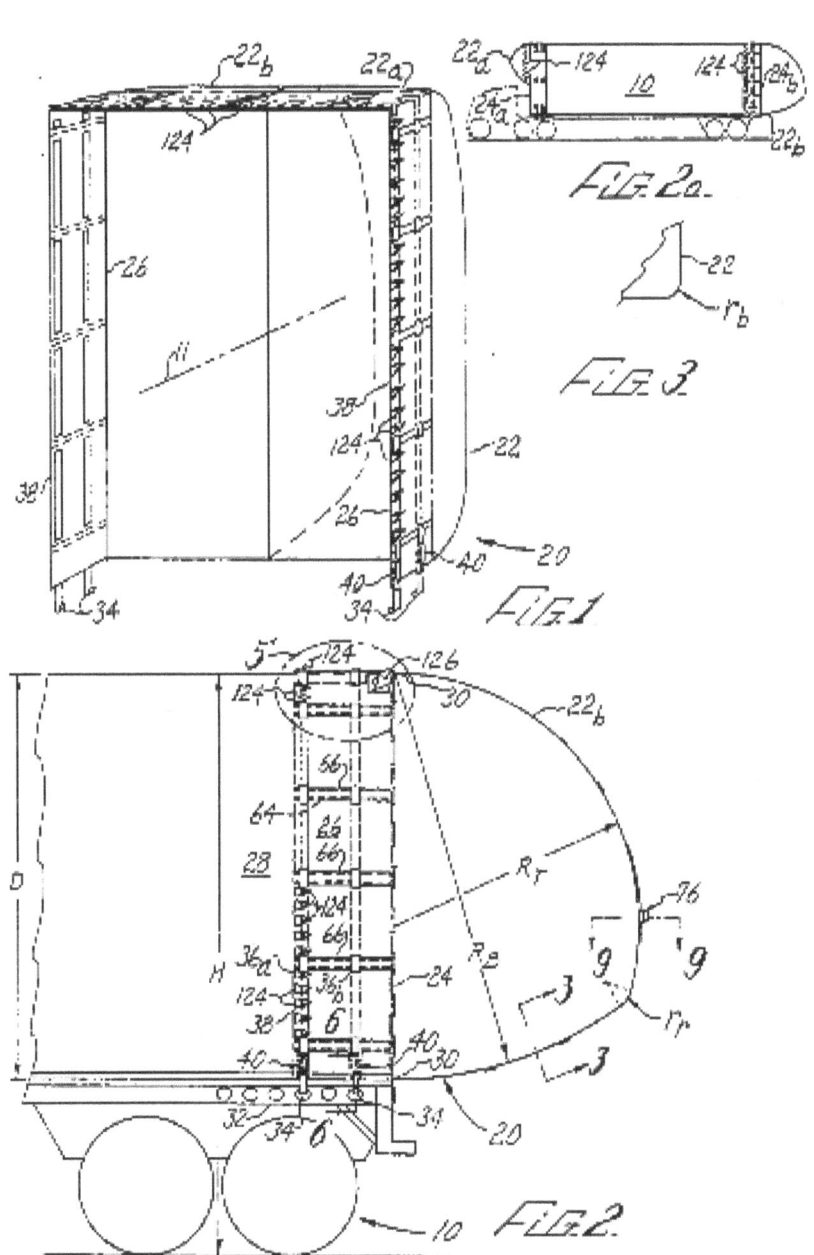

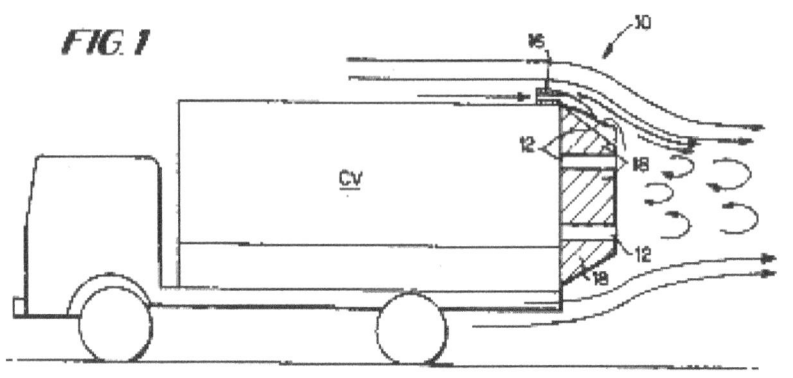

FIG. 1

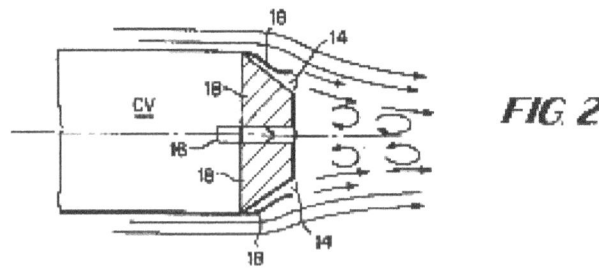

FIG. 2

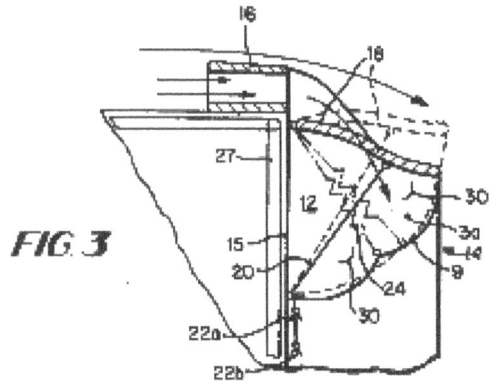

FIG. 3

FIG. 4

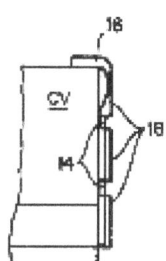

FIG. 5

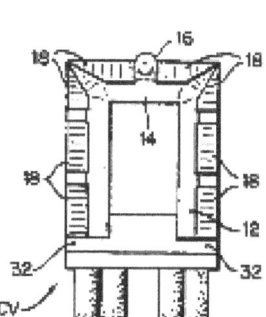

FIG. 6

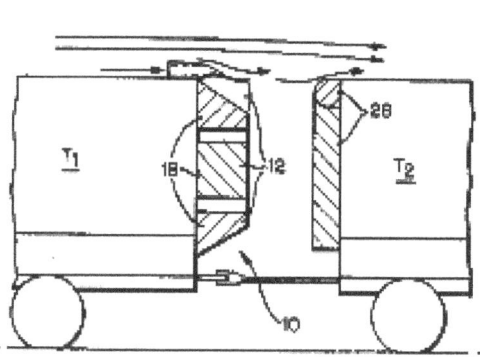

FIG. 7

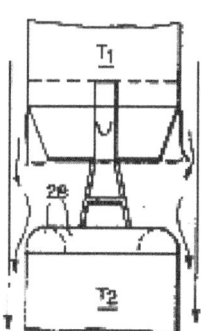

Nov. 7, 1944. A. BOYNTON 2,361,924
ATMOSPHERIC RESISTANCE EQUALIZING MEANS FOR MOVING VEHICLES
Filed April 20, 1943

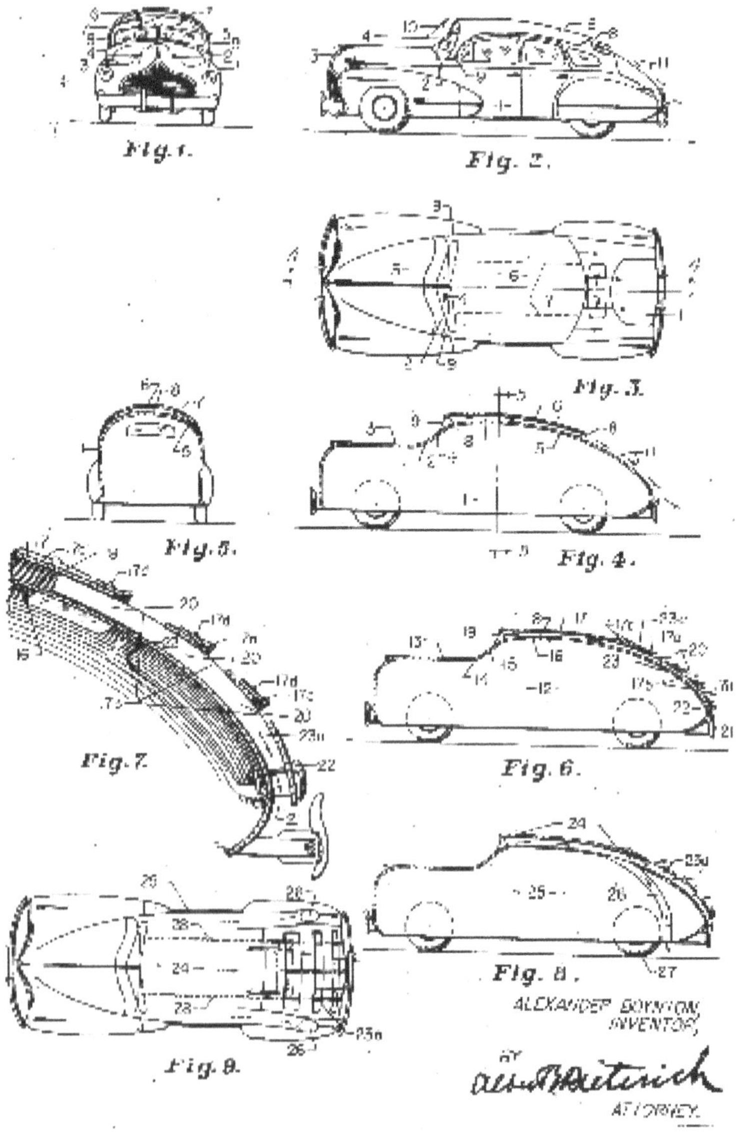

Fig. 1.
Fig. 2.
Fig. 3.
Fig. 5.
Fig. 4.
Fig. 7.
Fig. 6.
Fig. 8.
Fig. 9.

ALEXANDER BOYNTON
INVENTOR
ATTORNEY

Appendix B

United States Patent: 4,343,506

August 10, 1982

Low-drag ground vehicle particularly suited for use in safely transporting livestock

Abstract

A low-drag truck consisting of a tractor-trailer rig (10) characterized by a rounded forebody and a protective fairing (16) for the gap conventionally found to exist between the tractor and the trailer, particularly suited for establishing an attached flow of ambient air along the surfaces thereof, and a forward facing, ram air inlet and duct (24 and 22) and a plurality of submerged inlets (18) and outflow ports (20) communicating with the trailer (14) for continuously flushing heated gasses from the trailer as the rig is propelled at highway speeds.

Inventors: Saltzman; Edwin J. (North Edwards, CA)
Assignee: The United States of America as represented by the Administrator of the (Washington, DC)
Appl. No.: 06/175,453
Filed: August 5, 1980
Current U.S. Class: 296/24.31; 105/1.2; 244/53B; 296/180.2; 296/91
Current International Class: B62D 35/00 (20060101); B62D 035/00 ()
Field of Search: 296/1S, 91, 24C 105/2R, 2A 244/53B
References Cited [Referenced By]
U.S. Patent Documents

2612027	September 1952	McGan
4092044	May 1978	Hoffman
4142755	March 1979	Keedy
4174083	April 1977	Mohn
4199185	April 1980	Woolcock

Primary Examiner: Peters, Jr.; Joseph F.
Assistant Examiner: Carroll; John A.
Attorney, Agent or Firm: Brekke; Darrell G. Manning; John R.
Government Interests

ORIGIN OF THE INVENTION

The invention described herein was made in the performance of work by an employee of the United States Government and may be manufactured and used by or for the Government for Governmental purposes without the payment of any royalties thereon or therefore.

Claims

What is claimed is:

1. In combination with a low-drag ground vehicle including a streamlined forebody and an elongated, sub-

stantially closed cargo box, said forebody being suitably shaped for establishing an attached flow of ambient air along the surface of the upper and side walls of the box as the vehicle is propelled at highway speeds, the improvement comprising:

Means for continuously flushing gases from the box as the vehicle is propelled at highway speeds including a ram air inlet system with an intake flushly disposed on said forebody and a plurality of submerged inlets defined in the walls of the box for ingesting portions of the attached flow and delivering the ingested portions of the flow to the interior of the box, and a plurality of flow discharge ports defined in the walls of the box aft of the plurality of submerged inlets for discharging gases from the interior of the box to the attached flow.

2. An improvement as defined in claim 1 wherein the pressure within the interior of said box is continuously maintained at a value greater than one atmosphere as the vehicle is propelled at highway speeds.

3. An improvement as defined in claim 1 or 2 wherein said ground vehicle comprises a tractor and a trailer having a gap defined there between, and an articulated air shield projected rearwardly over the tractor enclosing said gap, said inlets are defined both in said shield and in the fore section of the trailer, and said flow discharge ports are defined in the mid and aft sections of said box.

4. An improvement as defined in claim 3 wherein said ram air inlet system has an intake opening disposed above said tractor and a duct with discharge opening communicating with the forward face of said box.

5. In combination with a low-drag ground vehicle characterized by a trailer including a closed, elongated flat-sided cargo box, a tractor with rounded side forebody edges connected to said trailer in spaced relation therewith, and an articulated fairing forming a rounded air shield projected rearwardly over the tractor protectively enclosing the space defined between the tractor and the trailer, said air shield and said forebody edges being of a streamlined configuration for establishing an attached flow of ambient air along the surfaces of the box as the vehicle is caused to progress at highway speeds, means for continuously flushing the atmosphere from the interior of said cargo box comprising:

A ram-air duct communicating with the forward face of the box and having an intake opening flushly disposed in said fairing above the tractor, a plurality of mutually spaced submerged inlets defined in the forebody of the box in communicating relation with the interior thereof for ingesting portions of an attached flow and delivering said portions of the flow to the interior of the box as the vehicle is operated at highway speeds, whereby an increased pressure condition is established within the box, and a plurality of out-flow ports for discharging gases from the interior of said box to the attached flow of ambient air at a rate sufficient to maintain the increased pressure condition established within the box.

6. A combination as defined in claim 5 further comprising submerged inlets defined in the fairing and connected in communicating relation with the space defined in the tractor and trailer for continuously circulating ingested portions as they flow through the space.

7. A combination as defined in claim 5 wherein each of said submerged inlets comprises a NACA submerged inlet characterized by divergent ramp walls and sharp edges.

Description

BACKGROUND OF THE INVENTION

1. Field of the Invention

The invention generally relates to the LDT (Low-Drag Truck) industry, and more particularly to an improved low-drag tractor-trailer rig or to a non-articulating straight truck characterized by a rounded forebody and a protective fairing for the gap between the tractor and the trailer, in the case of the former, for establishing an attached flow of ambient air extended along surfaces of the trailer and a forward facing ram air inlet and submerged inlets for continuously flushing heated gasses from the trailer as the rig is propelled at highway speeds.

2. Description of the Prior Art

The use of tractor-trailer rigs or combinations having rounded forebodies and fairings forming protective shields between the cabs and the trailers thereof in order to facilitate the attachment thereto of an air flow, for purposes of reducing aerodynamic drag, and consequently fuel consumption, generally is well known. For example, note the patent to Servais et al U.S. Pat. No. 4,036,519 which illustrates a tractor-trailer rig utilizing an LDT-type forebody.

Additionally, the use of vents for flushing the atmosphere of such trailers also is well known. For example, see the patent to Stone U.S. Pat. No. 4,018,480 which discloses a trailer having a plurality of ventilation openings formed through the wall panels near the lower ends thereof.

Notwithstanding the fact that the use of LDT vehicles generally is well known, unique problems arise, particularly in the livestock industry, when a use of such vehicles is contemplated. One of the problems of particular significance is that of the intense build-up of heat which can be expected within the enclosure or fairing provided for the gap between the cab and trailer. This build-up of heat tends to create a condition normally tragic for livestock riding near the front of the trailer.

Moreover, because of the nature of the cargo transported by livestock haulers, toxic fumes and gasses accumulate in stagnant pockets within the cargo box leading to discomfort and even physical damage to the cargo. Such a result is obtained even when employing typical livestock haulers having ventilation ports and the like defined in the walls of the cargo box.

It is perhaps appropriate to note the combination of over-heat, generation of toxic gasses within the cargo box, as well as the uneven ventilation involved, often results in consequences more drastic than simply the comfort of livestock, though the comfort factor is in itself important. During a year when livestock losses associated with shipping fever were tabulated, 1974, the total losses were on the order of $500 million dollars. Such losses are indeed important to both producers and consumers.

It is therefore the general purpose of the instant invention to provide an improved, LDT or low-drag ground vehicle, either of the non-articulating type or an articulating tractor-trailer rig, particularly suited for use in transporting livestock and other cargo requiring continuous temperature and atmosphere control.

OBJECTS AND SUMMARY OF THE INVENTION

It is an object of the instant invention to provide an improved, low-drag ground vehicle either of the non-articulating type or the articulating type.

It is another object to provide in combination with an LDT means for continuously controlling temperature and flushing toxic atmosphere from the interior of the cargo box or trailer thereof.

It is another object to provide in combination with a low-drag tractor-trailer rig a forward facing ram air inlet and a plurality of submerged inlets particularly adapted for ingesting portions of an attached flow of air and introducing the ingested portions of the flow into the trailer for thereby continuously flushing the atmosphere there from without significantly enhancing vehicle drag.

These and other objects and advantages are achieved through the use of a forward facing, ram air inlet and a plurality of submerged inlets interconnected in combination with the fairing and trailer of a low-drag tractor-trailer combination or rig, whereby the trailer is continuously pressurized through an ingestion of portions of an attached flow of air established as the vehicle is propelled at highway speeds, as will become more readily apparent by reference to the following description and claims in light of the accompanying drawings.

DESCRIPTION OF THE DRAWINGS

FIG. 1 is a perspective view of an LDT (Low-Drag Truck) embodying the principles of the instant invention.

FIG. 2 is a top plan view of a conventional base line cab-over vehicle including arrows diagrammatically illustrating the aerodynamic turbulence associated with the exterior surfaces and the stagnation of atmosphere within the cargo box thereof.

FIG. 3 is a top plan view of an operating LDT including arrows diagrammatically illustrating an attached air flow established along the exterior surfaces and the flushing effects achieved employing the forward facing, ram air inlet and the several submerged inlets, in the manner consistent with the principles of the instant invention.

FIG. 4 is a rear elevational view of the LDT shown in FIG. 1.

FIG. 5 is a side-elevational view of the LDT shown in FIG. 3 diagrammatically depicting ingestion of portions of the attached air flow.

FIG. 6 is a fragmented, cross-sectional view of a submerged inlet comprising a so-called NACA submerged inlet depicting vortexes generated by sharp edges provided for enhancing the mass/flow ratio of air ingested by the duct.

DESCRIPTION OF THE PREFERRED EMBODIMENT

Referring now to the drawings, with more particularity, wherein like reference characters designate like or corresponding parts throughout the several views, there is shown in FIG. 1 a low-drag ground vehicle, generally designated 10, conforming in its configuration to that of a well known Low-Drag Truck, herein referred to simply as an LDT.

It is here observed that as illustrated in FIG. 2, where a rig, such as the conventional base-line, cab-over truck is propelled at highway speeds, excessive vortexing and turbulence occurs in the resultant air stream. The operation of such vehicles is, therefore, necessarily attended by excessive vehicle drag; an attachment of stagnant pools of "highway gasses", including toxic fumes; and generally poor vehicle ventilation.

As shown in FIGS. 1, 3 and 5, the vehicle 10 comprises an LDT including a tractor 12, a trailer 14, and an articulated fairing 16 projected rearwardly over the cab to the trailer, in protective relation with the resultant gap normally found to exist between the cab and the trailer. Since the purpose of and function of an LDT generally are well known, a detailed description of the vehicle 10 is deemed unnecessary to provide for a complete understanding of the invention. Therefore, it is believed sufficient to appreciate the fact that, as illustrated in FIG. 3, the vehicle 10 is provided with a rounded forebody which, in cooperation with the fairing 16, serves to establish an air stream about the exterior of the vehicle which comprises a flow aerodynamically attached to the surfaces of the vehicle, as the vehicle is propelled at highway speeds.

It is particularly important to note that the vehicle 10 includes a ram air inlet and a plurality of submerged inlets 18, the purpose of which is to ingest portions of the impinging and attached air flow and introduce the ingested portions of the air flow into the interior of the fore section of the trailer 14, while out-flow ports 20 are provided at the mid and rear sections of the trailer for accommodating a discharge of gasses from the trailer. Thus a continuous flushing of the atmosphere from the trailer is achieved in response to the vehicle 10 being operated at highway speeds.

It is believed important to note that the term "submerged inlets", as herein employed, refers to inlets which are of a design frequently referred to as NACA submerged inlets inasmuch as these inlets or ducts were developed by the National Advisory Committee for Aeronautics in order to achieve efficient ingestion of air from the surface of the fuselage of aircraft for delivery to engines, equipment, components, and the like.

Even though the details of ram air inlets and the submerged inlets 18 are well known and form no part of the claimed invention, it should be noted that a use of these inlets is deemed critical to a satisfactory operation of the invention as herein described. This results from the fact that these inlets comprise ducts which are so designed as to obtain optimum delivery of air with a minimal drag penalty being imposed. A NACA submerged inlet is, as illustrated in FIG. 6, provided with a ramp and a pair of curved, divergent walls, which intersect the surface to form sharp edges. The sharp edges defined by the walls and the surface produce vortexes which, in combination with the divergence thereof, re-energize and thin a boundary layer as it develops along the ramp, whereby the boundary layer is controlled and the efficiency of the inlet is enhanced. Therefore, it is particularly important to appreciate that the submerged inlets 18 have a unique capability of ingesting optimum quantities of air from an attached flow, all without impairing the characteristics of the flow as it passes over the exterior surfaces of the LDT or vehicle 10.

As further illustrated in FIG. 3, ingested portions of the air flow are introduced in the forward portions of the trailer 14 while the generally toxic atmosphere including gasses generated within the trailer are swept, by these portions of the flow, and are discharged from the trailer at the mid and rear portions thereof, with a resultant minimal attendant stagnation or pooling of the atmosphere within the trailer.

As a practical matter, while the number, shape and distribution of the out-flow ports 20 may be established empiracally, and are varied in proportion to the mass flow rate of air to be ingested, the number and distribution of the inlets 18 and out-flow ports 20 are such as to assure a continuous pressurization of the trailer 14 as the vehicle 10 is operated at highway speeds.

As also shown in FIG. 3, submerged inlets 18 also are connected in communication with the space between the cab and trailer for flushing downwardly air and heat entrapped in this space. Thus the space is continuously cooled.

A ram air duct 22 will be provided for achieving a further introduction of the air flow to the trailer. The duct 22, as shown, includes an inlet orifice 24, FIG. 3, and a discharge orifice 26 communicating with the interior of the trailer 14, preferably through the leading end wall thereof. The ram air duct 22 is of a suitable design, the details of which form no part of the claimed invention, however, the duct 22 preferably is comprised of a flexible duct which accommodates articulation of the tractor-trailer rig. Additionally, the inlet orifice 24 is provided with smooth, nicely rounded edges, the radius of which preferably is on the order of six inches. Consequently, ram air is ingested via the orifice 24 with minimal attendant turbulence.

Finally, while not shown, it should be apparent that the ducts 18 and 22 are particularly suited for establishing temperature control, particularly in colder climates, for the trailer 14, since the ducts readily may be provided with selectively operable closure members for purposes of restricting the flow of air therethrough.

OPERATION

It is believed that in view of the foregoing description, the operation of the instant invention is readily understood, however, for the sake of assuring a complete understanding, the operation thereof is at this point briefly reviewed.

With the vehicle 10 assembled in the manner hereinbefore described, the vehicle 10 is prepared for operation, such as the hauling of livestock and the like. However, it is to be understood that the utility of the vehicle is not limited to the field of transporting livestock but may be employed in any field in which it is desired that an LDT be utilized for hauling cargo requiring improved ventilation and/or temperature control.

With reference to FIG. 3, it can be seen that as the vehicle 10 is driven at highway speeds there is established along the exterior surfaces thereof an attached air flow. The submerged inlets 18 serve to ingest portions of the attached air flow and introduce the portions into the interior of the trailer 14. Additionally, ram air is introduced into the trailer 14 via the orifice 24 and the duct 22. Of course, the space between the cab and trailer continuously is flushed as air moves from the submerged inlets 18 formed in the fairing 16 through the space to be discharged therebeneath. Thus an increased atmospheric pressure is established within the trailer 14 as well as the space between the cab and trailer.

Simultaneous with the establishment of increased pressures within the trailer 14, the outflow ports 20 accommodate a discharge of gasses from the trailer to the region behing the vehicle and to the air stream flowing along the sides of the trailer, all without significantly enhancing turbulence, for thus disrupting the attached air-flow, and inducing drag. It should be recognized that the portion of air flowing out of ports at the base, or rear surface, of the cargo box will reduce the drag of the vehicle. This flushing or discharge of gasses is, in operation, continuous for thus continuously renewing the air and controlling the temperature within the trailer.

In view of the foregoing, it is believed to be readily apparent that through the use of a so-called LDT design for livestock haulers, fuel economy is realized, and moreover, through the use of ram air and submerged inlets, the life supporting environment within the trailer 14 is greatly enhanced. Thus there has been provided a practical solution to many of the problems heretofore plaguing designers of livestock haulers and the like.

Appendix C

U.S. Patent number: 6,892,989
May 17, 2005
Inventors: Whitmore; Stephen A. (Lake Hughes, CA), Saltzman; Edwin J. (North Edwards, CA), Moes; Timothy R. (Lancaster, CA), Iliff; Kenneth W. (Lancaster, CA)
Assignee: The United States of America as represented by the Administrator of the National Aeronautics and Space Administration (Washington, DC)

Abstract

A method for reducing drag upon a blunt-based vehicle by adaptively increasing forebody roughness to increase drag at the roughened area of the forebody, which results in a decrease in drag at the base of this vehicle, and in total vehicle drag.

U.S. Patent Documents

Number	Date	Inventor
2261558	November 1941	Orloff
2800291	July 1957	Stephens
3319593	May 1967	Papst
4750693	June 1988	Lobert et al.
4907765	March 1990	Hirschel et al.
5114099	May 1992	Gao
5133516	July 1992	Marentic et al.
5133519	July 1992	Falco
5171623	December 1992	Yee
5263667	November 1993	Horstman
5346745	September 1994	Banyopadhyay
5378524	January 1995	Blood
5386955	February 1995	Savill
5505409	April 1996	Wells et al.
5598990	February 1997	Farokhi et al.
5618215	April 1997	Glydon
5803410	September 1998	Hwang
5836016	November 1998	Jacobs et al.
5848769	December 1998	Fronek et al.
6612524	September 2003	Billman et al.

Other References

NASA Article "A Base Drag Reduction Experiment on the X-33 Linear Aerospike SR-71 Experiment (LASRE) Flight Program," dated Mar. 1999.

Claims

What is claimed is:

1. A method for reducing the drag of a vehicle having a forebody and a base, comprising irregularly coarsening a surface of the vehicle in order to increase drag along the coarsened surface, thereby reducing drag aft of the coarsened surface, whereby approximately 1/3 of the forebody of the vehicle is coarsened, wherein the vehicle is a flight vehicle, whereby the coarsening is accomplished by attaching a coarsening agent to the vehicle surface, the coarsening agent being applied to approximately 1/3 of the forebody of the vehicle, the coarsening agent having an average diameter of approximately 0.035 inches, whereby the coarsening agent is suspended in paint, wherein the coarsened surface has an equivalent sand-grain roughness of between approximately 0.02 and 0.05 inches, whereby MEMS (Micro-Electro-Mechanical Systems) controllers are used to adaptively vary the coarsened surface equivalent roughness according to Mach number.

2. A method for reducing the drag of a flight vehicle having a forebody wetted area and a base, comprising coarsening approximately 1/3 of the forebody wetted area with a coarsening agent having an average approximate diameter of 0.035 inches to create a coarsened surface, such that the coarsened surface has an equivalent sand-grain roughness of between approximately 0.02 and 0.05 inches, and the coarsened surface includes MEMS (Micro-Electro-Mechanical Systems) controllers to adaptively vary the coarsened surface equivalent roughness according to Mach number in order to increase drag along the coarsened surface, thereby reducing drag along the base.
Description

BACKGROUND OF THE INVENTION

1. Field of the Invention

This invention relates to methods and devices for reducing drag on blunt-body vehicles.

2. Description of the Related Art

Current proposed shapes for single-stage-to-orbit vehicles like the Lockheed-Martin X-33 and "Venture-Star" reusable launch vehicle (RLV) have extremely large base areas when compared to previous hypersonic vehicle designs. As a result, base drag, especially in the transonic flight regime, is expected to be very large, and will likely dominate or overwhelm all other factors relevant to the vehicle performance. Excessive base drag could seriously limit the range of available landing sites for the "Venture Star" and will reduce payload capability. The unique configuration of Lockheed-Martin RLV with its very large base are and relatively low forebody drag, offers the potential for a large increase in overall vehicle performance, if the base drag can be reduced significantly.

There have been previous attempts to generally address the issue of drag reduction by altering the surface of a vehicle.

U.S. Pat. No. 4,907,765 discloses a wall having a drag-reducing configuration comprising a wall structure with sharp edged ridges separated by valleys that have drag reducing characteristics.

U.S. Pat. No. 5,378,524 discloses a vehicle with an outer surface that includes a matrix of cavities. The vehicle is selected from the group consisting of automobile, airplane and boat.

U.S. Pat. No. 5,346,745 discloses a plurality of surface elements arranged in rows on the surface of an object, with the surface elements of each row being arranged generally orthogonal to the direction of relative motion of the object. Each surface element includes means defining a cavity, and the cavities are interconnected by means of passageways to facilitate fluid communication therebetween. The passageways facilitate equalization of pressure between the cavities of the surface elements in each row, which ultimately results in reducing turbulence around the object.

U.S. Pat. No. 2,261,558 discloses providing recesses of various sorts and shapes on the surface of a vehicle, such as will minimize the air and water resistance offered by the vehicle, especially when proceeding at relatively high speeds.

U.S. Pat. No. 5,171,623 discloses drag-reducing surface depressions that are shaped like sections of truncated cones, or hexagonal prisms, geodesic domes, and that cover the entire surface of the body of the vehicle.

What each of these prior art approaches has in common is that they use depressions in the vehicle surface to reduce drag at the locations of the depressions. However, none of these approaches address the issues peculiar to blunt-based vehicles with extremely large base areas, and none disclose reducing overall vehicle drag by increasing drag at particular areas of a vehicle. Therefore, a need exists for a method of reducing drag in a blunt-based vehicle with an extremely large base area that is effective, easy to implement, applicable to all types of blunt-based vehicles at all speeds, and does not decrease the inherent structural integrity of the vehicle

SUMMARY OF THE INVENTION

In view of the foregoing disadvantages inherent in the known types of drag reduction methods now present in the prior art, the present invention provides a new method of drag reduction wherein the same can be utilized for blunt-body vehicles with large base areas.

The general purpose of the present invention, which will be described subsequently in greater detail, is to provide a new drag reduction method which has many novel features that result in a method of reducing drag which is not anticipated, rendered obvious, suggested, or even implied by any of the prior art methods, either alone or in any combination thereof.

The methods discussed in this document offer a means to achieve such reductions. The method includes reducing the drag of a vehicle having a forebody and a base by coarsening the surface of the vehicle in order to increase drag along the coarsened surface, thereby reducing drag aft of the coarsened surface.

BRIEF DESCRIPTION OF THE DRAWINGS

FIG. 1 is a graph of the subsonic correlation of base and viscous forebody drag coefficients.

FIG. 2 is a visualization of a base pumping mechanism.

FIG. 3 is a graph depicting the visualization of the "drag bucket."

FIG. 4 depicts the layout of the LASRE forebody grit.

FIG. 5 is a rear view of an exemplary vehicle where the coarsened surface is depicted by the shaded area.

FIG. 6 is a perspective view of an exemplary vehicle where the coarsened surface is depicted by the shaded area.

DESCRIPTION OF THE PREFERRED EMBODIMENT(S)

The detailed description set forth below in connection with the appended drawings is intended as a description of presently preferred embodiments of the invention and is not intended to represent the only forms in which the present invention may be constructed and/or utilized. The description sets forth the functions and the sequence of steps for constructing and operating the invention in connection with the illustrated embodiments. However, it is to be understood that the same or equivalent functions and sequences may be accomplished by different embodiments that are also intended to be encompassed within the spirit and scope of the invention.

Drag reduction tests were conducted on the LASRE/X-33 flight experiment, a roughly 20percent scale model of an X-33 forebody with a single aerospike engine at the rear. The experiment apparatus was mounted on top of an SR-71 aircraft. The tests investigated a novel method for reducing base drag by adding surface roughness along the LASRE forebody. Calculations showed a potential for base drag reductions of 8-14 percent. Flight results corroborate the base drag reduction, with actual reductions of 15 percent in the high-subsonic flight regime. An unexpected positive result of the test was that drag reductions persist well into the supersonic flight regime. This result is extremely important because it demonstrates that the boundary layer still has a significant influence on the base separation, even in the presence of oblique shock waves and supersonic expansion waves, i.e. the base area does not "shock-off" from the rest of the external flow field.

For blunt-based objects whose base areas are heavily separated, i.e. experience detached flow conditions, a clear relationship between the base drag and the "viscous" forebody drag has been demonstrated. This trend is presented in FIG. 1 along with subsonic LASRE drag data. The trend presented in FIG. 1 shows that as the forebody drag is increased; generally the base drag of the projectile tends to decrease.

This base-drag reduction is a result of boundary layer effects at the base of the vehicle. The shear layer caused by the external flow rubbing against the separated air in the base region act as a jet pump and serves to reduce the pressure in the base area. This pumping effect is graphically illustrated in FIG. 2. The viscous high-speed external flow "pulls" air out of the base region because of 1) viscous shear forces in the shear layer and 2) the low static pressure in the external flow according to Bernoulli principles. These two effects cause the air to be "pumped" away from the base and the pressure to be reduced in the base region. Reduced pressure results in increased base drag.

The surface boundary layer acts as an "insulator" between the external flow and the air at the base. Consequently, a thicker boundary layer reduces the two base-drag causing effects. As the forebody drag is increased, the boundary layer thickens at the aft end of the forebody thereby reducing the effectiveness of the pumping mechanism and resulting in reduced base drag.

Because the LASRE drag data lie on the steep, nearly vertical, portion of the curve, a result of the large base drag, a small increment in the forebody friction drag should result in a relatively large decrease in the base drag. Conceptually, if the added increment in forebody skin drag is optimized with respect to the base drag reduction, then it is possible to reduce the overall drag of the configuration.

In order to predict the expected magnitudes of these drag reductions, a mathematical model of the LASRE base drag coefficient, which has the viscous forebody drag coefficient as a parameter was developed. The model accounts for flow compressibility using relationships defined by the Karman-Tsien correction and rules of similarity for transonic flow. If one plots the total drag of the vehicle as a function of the forebody drag, then a minimum value or "drag bucket" will occur at some value for the forebody drag coefficient. The model predictions are plotted in FIG. 3 along with measured data for several hypersonic lifting-body and wing-body configurations: X-15, M2-F1, M2-F2, Shuttle, HL-10, X-24A, X24-B and the LASRE (taken to represent the characteristics of the X-33/Venture-Star). Whereas most of the previously flown hypersonic shapes lie near or slightly to the right of the drag minimum, the X-33 lies far to the left of the drag minimum, as shown in FIG. 3. This behavior is a result of the previously discussed large "base-to-wetted-area" ratio. Thus the X-33 RLV shape offers a potentially high pay-off for overall vehicle drag reduction by simply increasing the vehicle forebody drag. The desired increase in forebody drag may be afforded by incorporating the roughness design into the surface thermal protection system (TPS).

The LASRE drag reduction experiment sought to verify the above hypothesis. In this experiment the boundary layer at the back end of the LASRE model was modified by increasing the forebody skin friction. Clearly, one of the most convenient methods of increasing the forebody skin drag is to add roughness to the surface.

Other methods such as using vortex generators to energize the boundary layer would probably work more effectively, however they must be used with care because their intrusiveness into the flow reduces their effectiveness with respect to the hypersonic re-entry vehicle problem. Benefits of using surface roughness are non-intrusiveness (minimal heating), small weight penalty, mechanical simplicity, and low cost. The resulting grit layout for the LASRE flight tests is depicted in FIG. 4. The resulting base drag reduction is compared against the predicted drag savings in FIG. 5.

Referring to FIGS. 5 and 6, the present invention comprises a method of reducing the drag of a vehicle 51 having a forebody 53 and a base 55, that includes coarsening the surface of the vehicle in order to increase drag along the coarsened surface 57, thereby reducing drag along base 55.

In one embodiment of the invention, approximately 1/3 of the forebody 53 of vehicle 51, or of the forebody wetted area, is coarsened, as shown if FIGS. 5 and 6. The coarsening may be accomplished by treating the surface of the vehicle with an abrasive, or attaching a coarsening agent to the vehicle. For example, the coarsening agent may be attached to a vehicle with an adhesive, then sealed with paint. Alternatively, the coarsening agent may be suspended in paint, such as a white enamel paint. The coarsening agent preferably has an average diameter of approximately 0.035 in., although this may vary from 0.001 in. to 0.10 in., depending on the specific characteristics of the subject vehicle. Ideally, but not necessarily, the resulting coarsened surface should have an equivalent sand-grain roughness of between approximately 0.02 and 0.05 in.

In an alternative embodiment, microelectromechanical systems (MEMS) controllers my be disposed beneath the coarsened surface, or within the coarsening mixture to adaptively vary the coarsened surface 57 equivalent roughness according to Mach number, or forward vehicle speed.

In another alternative embodiment, vortex generators may be used in the case of lower-speed vehicles. Vortex generators are very efficient devices for increasing forebody drag and energizing the boundary layer, but are preferably used on vehicles operating at supersonic speeds and below.

All embodiments of the invention may be practiced on flight, ground or aquatic vehicles, and at all speeds, to include hypersonic, although certain modifications known to those skilled in the art may be necessary for a particular application.

While the present invention has been described with regards to particular embodiments, it is recognized that additional variations of the present invention may be devised without departing from the inventive concept.

Bibliography

Primary Sources

Anderson, J. W., J.C. Firey, P. W. Ford and W. C. Kieling, "Truck Drag Components by Road Test Measurement." Society of Automotive Engineers Transactions (1965), vol. 73,

Carr, G. W. "The Aerodynamics of Basic Shapes for Road Vehicles, part 1 — Simple Rectangular Bodies, Report no. 1968/2." Motor Industry Research Association November 1967.

Comprehensive Truck Size and Weight Study. (Washington, DC: US DOT FHA), in four volumes, August 2000. http://www.fhwa.dot.gov/policy/otps/truck/finalreport.htm, accessed 3 June 2009.

Dempsey, E.A. "Vehicle Body and Attachment Therefor." [sic] U.S. Patent No. 2,514,695 of July 1950.

Diebler, Corey and Mark Smith. A Ground-Based Research Vehicle for Base Drag Studies at Subsonic Speeds. Edwards, CA: NASA TM 2002-210737, November 2002.

FHWA Bridge Formula, http://training.ce.washington.edu/WSDOT/Modules/04_design_parameters/bridge_formula.htm, accessed 31 December 2009.

Favre, Alexandre. "Aircraft Wing Flap with Leading Edge Roller." U.S. Patent No. 2,569,983, 2 October 1951.

Gross, Donald S. Wind Tunnel Tests of Trailmobile Trailers. 3rd series / prepared by Donald S. Gross, College Park, MD: University of Maryland, College of Engineering, Glenn L. Martin Institute of Technology, Wind Tunnel Operations Dept., [1955?].

Flynn, Harold and Peter Kyropoulos. "Truck Aerodynamics." SAE Transactions 1962 vol. 70: 297-308.

Hoerner, Sighard F. Aerodynamic Drag: Practical Data on Aerodynamic Drag, Evaluated and Presented by Sighard F. Hoerner. Midland Park, NJ, 1951.

Hoffman J. A., D. R. Sandlin. A Preliminary Investigation of the Drag and Ventilation Characteristics of Livestock Haulers. Edwards, CA: NASA CR 170408, 1983..

Horton, V. W., R. C. Eldgredge, and R. E. Klein. Flight-Determined Low-Speed Lift and Drag Characteristics of the Lightweight M2-F1. Edwards, CA: NASA TN-D3021, 1965.

Keedy, Edgar L. "Vehicle Drag Reducer." U.S. Patent No. 4,142,755, 6 March 1979.

Labé, François. "Drag Reducer for Rear End of Vehicle." U.S. Patent No. 4,978, 162, Dec. 18, 1990.

Landman, Drew, Richard Wood, Whitney Seay and John Bledsoe. Understanding Practical Limits to Heavy Truck Drag Reduction. #2009-01-2890. Society of Automotive Engineers. Warrendale, PA, 2009.

Lissaman, P. B. S. "Development of Devices to Reduce the Aerodynamic Resistance of Trucks." Society of Automotive Engineers Annual Meeting, Seattle, WA, 11-14 August 1975.

Meyers, Cornelius T. "Air-Deflecting Device." U.S. Patent No. 1,352,679.

Bibliography

Meyers, Dale D. "The Origins of the Space Shuttle." Jeffrey Hoffman. 16.885J Aircraft Systems Engineering, Fall 2005. Massachusetts Institute of Technology: MIT OpenCourseWare, http://ocw.mit.edu. License: Creative Commons BY-NC-SA. Lecture 1, accessed 26 May 2011.

Moes, Tim, Tony Whitmore, Ken Iliff, and Ed Saltzman. "Base Drag Research Using a Ground Research Vehicle." October 12, 2000.

Montoya, Lawrence C., and Louis L. Steers. Aerodynamic Drag Reduction Tests on a Full-Scale Tractor-Trailer Combination with Several Add-On Devices. Edwards, CA: NASA TM-X 56028, 1974.

_____."Aerodynamic Drag Reduction Tests on a Full-Scale Tractor Trailer Combination with Several Add-On Devices." (n.d.) NASA Flight Research Center and the DOT Transportation Systems Center.

Muirhead, Vincent U. An Investigation of Drag Reduction for Tractor-Trailer Vehicles. Edwards, CA: NASA CR 144877, October 1978.

_____. An Investigation of the Internal and External Aerodynamics of Cattle Trucks. Edwards, CA: NASA CR 170400, 1983.

_____. An Investigation of Drag Reduction on Box-Shaped Ground Vehicles. Lawrence, KS: University of Kansas Center for Research, Inc., 1976.

_____. Final Report on An Investigation of Drag Reduction on Box-Shaped Ground Vehicles. Lawrence: The University of Kansas, 1976. KU-FRL 180.

Muirhead, Vincent U., and Edwin J. Saltzman. "Reduction of Aerodynamic Drag and Fuel Consumption for Tractor-Trailers Vehicles." Journal of Energy. vol. 3, no. 5 September-October 1979.

NASA Technology Note: Energy Efficient Livestock Hauler with Improved Ventilation Temperature Control. NASA Ames Research Center (n.d.).

Petersen, Randall L. Drag Reduction Obtained by the Addition of a Boattail to a Box Shaped Vehicle. Edwards, CA: NASA CR 163113, August 1981.

Potter, R. D. "Inflatable Streamlining Apparatus for Vehicle Bodies." U.S. Patent No. 2,737,411, 6 March 1956.

Saltzman, Edwin J. "A Summary of NASA Dryden's Truck Aerodynamic Research." 821284. Society of Automotive Engineers. Warrendale, PA, 1982.

_____. "Low-drag ground." U. S. Patent No. 4,343,506, 10 August 1982.

Saltzman, E. J., and R. R. Meyer, Jr., Drag Reduction Obtained by Rounding Vertical Corners on a Box-Shaped Ground Vehicle. Edwards, CA: NASA-TM-X-56023, 1974.

Saltzman, Edwin J., and Robert M. Meyers. A Reassessment of Heavy-Duty Truck Aerodynamic Design Features and Priorities. Edwards, CA: NASA/TP 1999 206574.

Saltzman, Edwin J., K. Charles Wang, and Kenneth W. Illif. "Flight-Determined Subsonic Lift and Drag Characteristics of Seven Lifting-Body and Wing-Truncated Reentry Vehicle Configurations with Truncated Bodies." American Institute of Aeronautics and Astronautics. 99-0383, 1999.

Saltzman, Edwin J., and Robert R. Meyer Jr., A Reassessment of Heavy-Duty Truck Aerodynamic Design Features and Priorities. NASA/TP-1999-206574, June 1999.

Saltzman, Edwin J., Robert R. Meyer and David F. Lux, Drag Reduction Obtained by Modifying a Box-Shaped Ground Vehicle. Edwards, CA: NASA, TM X-56027 October 1974.

Saunders, Walter Selden. "Drag Reducer for Land Vehicles." U.S. Patent No. 3,397,120, 10 October 1972.

Schlichting, H. "Aerodynamic Problems of Motor Cars." AGARD Report 307. October 1960.

Servais, R. A. "An Experimental and Analytical Investigation of Truck Aerodynamics." Proceedings of the Conference/Workshop on Reduction of Aerodynamic Drag of Trucks. Washington, D.C.: National Science Foundation, October 1974.

Servais, Ronald A. "Streamlining Apparatus for Articulated Road Vehicle." U.S. Patent No. 3,945,677, 23 March 1976.

Sheridan, Arthur E., and Steven J. Grier, Drag Reduction Obtained by Modifying a Standard Truck. NASA TM-72846, 1978.

Sherwood, A. Wiley. University of Maryland Wind Tunnel Report No. 85: Wind Tunnel Test of Trailmobile Trailers. University of Maryland Wind Tunnel Department: College Park, MD, 1953.

Smith, Gary L. "Commercial Vehicle Performance and Fuel Economy." Society of Automotive Engineers. SP-355, January 1970.

Smith, William Laughton. The Pretensions of Thomas Jefferson to the Presidency Examine. (n.p., 1796) In Richard Hofstadter, Anti-intellectualism in American Life New York: Alfred A. Knopf, 1963. 5th edition. Reprint, 1969.

Stalker, E. A. "Means of Reducing the Fluid Resistance of Propelled Vehicles." U.S. Patent No. 2,037,942, April 1936.

Stamm, A. F. "Tractor-Trailer Airstream Control Kit." U.S. Patent No. 2,863,695, 9 December 1958.

Steers, Louis L., and Edwin J. Saltzman. "Reduced Truck Fuel Consumption Through Aerodynamic Design." Journal of Energy. vol. 1, no 1. September-October 1977.

Steers, Louis L., and Lawrence C. Montoya. Study of Aerodynamic Drag Reduction on a Full-Scale Tractor-Trailer. National Aeronautics and Space Administration, Dryden Flight Research Center, Edwards, CA, in conjunction with the US Department of Transportation, Washington, D.C. DOT-TSC-OST-76-13. National Technical Systems Service, Springfield, Va., 1976.

Steers, L. L., and L. C. Montoya. Study of Aerodynamic Drag Reduction on a Full Scale Tractor-Trailer. Report No. DOT-TSC-OST-76-13. Washington, D. C.: U.S. Department of Transportation, April 1976.

Steers, Louis L., and Edwin J. Saltzman. "Reduced Truck Fuel Consumption through Aerodynamic Design." Journal of Energy vol. 1, No 5. September-October 1977.

Van Dyke, Milton. An Album of Fluid Motion. Stanford, CA: The Parabolic Press, 1982. Reprint 2007.

Bibliography

Werlé, Henri. Onera Aerospace Laboratory.

Whelan, William. "Air Drag Reduction Unit for Vehicle." U.S. Patent No. 6,092,861, July 25, 2000.

"White Paper on Fuel Economy." Kenworth Truck Company. October 2001.

Whitmore, Stephen A., and Timothy R. Moes, A Base Drag Reduction Experiment on the X-33 Linear Aerospike SR-71 (LASRE) Flight Program. Edwards, CA: NASA TM 1999-206575, 2000.

Whitten, W. David. "Collapsible Streamlined Tail for Truck and Trailers." U.S. Patent No. 6,309,010 B1, Oct. 30, 2001.

Wiley, N.C. "Demonstration Aerodynamic Drag Reduction for the purpose of Reducing Fuel Consumption of Trucks." Stanford, CT, 5 December 1974.

Wood, Richard M., and Steven X. S. Bauer. "Simple and Low-Cost Aerodynamic Drag Reduction Devices for Tractor-Trailer Trucks." Society of Automotive Engineers. 2003-01-3377.

U.S. Patents 6799791; 3,371,146; 3,929,369; 3,999,797; 4,095,835; 4,102,548; 4,131,309; 4,142,755; 4,316,630; 4,458,936; 4,601,508; 4,702,509; 4,978,162; 5,058,945; 5,108,145; 6,092,861; 6,309,010; 6,409,252; 6,742,616; 6,854,788; 7,216,923, and 4,095,835. http://www.google.com/patents?id=thGqAAAAEBAJ&dq=6467833, accessed 3 June 2009.

Secondary Sources

Aitken, Hugh G. J. Scientific Management in Action: Taylorism at Watertown Arsenal, 1908-1915. Princeton: Princeton University Press, 1985.

Allen, John E. Aerodynamics: The Science of Air in Motion. London: Hutchinson & Co., Ltd, 1963. 2nd edition, Granada Publishing, 1982.

Angus Beef Bulletin, September 2001.

Anton, Philip S., Eugene C. Gritton, Richard Mesic, Paul Steinberg, Dana J. Johnson, Michael Block, Michael Scott Brown, Jeffrey A. Drezner, James Dryden, Thomas Hamilton, Thor Hogan, Deborah Peetz, Raj Raman, Joe Strong, William P. G. Trimble. Wind Tunnel and Propulsion Test Facilities: An Assessment of NASA's Capabilities to Serve National Needs. Santa Monica, CA: RAND Corporation. 2004.

Baida, Peter. Poor Richard's Legacy: American Business Values from Benjamin Franklin to Donald Trump. New York: William Morrow and Company, Inc., 1990.

Bijker, Wiebe E., Thomas P. Hughes, and Trevor J. Pinch, editors. The Social Construction of Technological Systems: New Directions in the Sociology and History of Technology. Cambridge, MA: MIT Press, 1987.

"Bovine Respiratory Disease: A New Look at Causes and Signs of Disease," http://www.mycattle.com/health/updates/brd-nlac.cfm.htm, accessed 10 October 2005.

Bureau of Labor Statistics. Labor Force Statistics from the Current Population Survey, 2009. http://www.bls.gov/cps/tables.htm, accessed 3 August 2010.

Caro, Robert A. The Power Broker: Robert Moses and the Fall of New York, New York: Vintage Books, 1975.

Conway, Erik M. High-Speed Dreams: NASA and the Technopolitics of Supersonic Transportation, 1945–1999. Baltimore: Johns Hopkins University Press, 2005.

Cowen, Ruth Schwartz. More Work for Mother: The Ironies of Household Technologies, from the Open Hearth to the Microwave. New York: Basic Books, 1983.

Crackel, Laura. "Stretching The Limits." Overdrive http://www.etrucker.com/apps/news/article.asp?id=39038.htm, accessed 3 June 2009.

Currin, John F. and W. Dee Whittier. "Recognition and Treatment of Bovine Respiratory Disease Complex." Virginia-Maryland Regional College of Veterinary Medicine, Virginia Tech, Publication Number 400-008, posted August 2000.

"Do You Prefer Traditional or Aerodynamic Trucks?" Overdrive, June 2005. http://www.etrucker.com/apps/news/article.asp?id=47579, accessed 3 June 2009.

Davies, Ebenezer. American Scenes, and Christian Slavery: A recent Tour of Four Thousand Miles in the United States. London: John Snow, 1849.

Dunlavy, Coleen. Politics and Industrialization: Early Railroads in the United States and Prussia. Princeton: Princeton University Press, 1994.

Expedited Freight Information Center, http://www.expeditersonline.com/artman/publish/truck-aerodynamics.html, accessed 3 June 2009.

Family Safety Magazine Summer 1978, vol. 37, no. 2, 2.

"Fleets Install Aerodynamic Truck, Trailer Parts," Transport Topics On Line, March 29, 2010, http://www.ttnews.com/articles/basetemplate.aspx?storyid=24075, accessed 2 June 2010.

Gelzer, Christian. "The Quest for Speed: An American Virtue, 1825-1930." PhD dissertation, Auburn University, 1998.

Gies, Joseph. "The Genius of Oliver Evans." AmericanHeritge.com. http://www.americanheritage.com/articles/magazine/it/1990/2/1990_2_50.shtml, accessed 3 June 2010.

Goldstein, Edward. "Wind Tunnels: Don't Count them Out." Aerospace American April 2010, vol. 48, no. 4.

Gorn, Michael H. Expanding the Envelope: Flight Research at NACA and NASA. Lexington: he University of Kentucky Press, 2001.

Hallion, Richard P. and Michael H. Gorn. On the Frontier: Experimental Flight at NASA Dryden. Washington, D.C.: Smithsonian Institution, 2003

Hansen, James R. Engineer in Charge: A History of Langley Aeronautical Laboratory, 1917-1958, NASA SP-4305, 1987.

_____. Spaceflight Revolution: NASA Langley Research Center from Sputnik to Apollo. NASA SP-4308, 1995.

Bibliography

Haynes, Leland. SR-71 Blackbird. "Clarence L. "Kelly" Johnson." http://www.wvi.com/~sr71webmaster/kelly1.htm, accessed 28 December 2009.

Hindle, Brooke, and Steve Lubar. Engines of Change: The American Industrial Revolution, 1790-1860. Washington, D.C.: Smithsonian Institution Press, 1986.

Hofstadter, Richard. Anti-intellectualism in American Life. New York: Alfred Knopf, 1963. 5th edition. Reprint 1969.

http://www.autoblog.com/2007/01/02/radical-semis-by-luigi-colani/, accessed 11 July 2011.

http://www.geocities.com/thetropics/1608/page11.htm, accessed 9 September 2009.

http://www.mycattle.com/health/updates/brd-recognition.cfm, accessed 10 October 2005.

Huber, Richard M. The American Idea of Success. New York: McGraw-Hill Book Company, 1971.

Hurt, Jr, H. H. Aerodynamics for Naval Aviators. Reprint, Aviation Supplies & Academics, 1965.

Kasson, John. Civilizing the Machine: Technology and Republican Values in American, 1776-1900. New York: Grossman Publishers, 1976. Reprint, New York: Penguin Books, 1977.

"Kenworth's T600 – A Look Back at the Truck that Broke the Mold." Land Line Magazine 2005.

Kline, Ronald and Trevor Pinch. "The Social Construction of the Automobile in the Rural United States." In Merritt Roe Smith and Gregory Clancey, Major Problems in the History of American Technology. Boston: Houghton Mifflin Company, 1998.

"Mack Research Shows Potential 8% Fuel Economy Improvement." 15 November 2006. http://www.truckinginfo.com/news/newsdetail.asp?news_id=57555&news_category_id=20.htm, accessed 11 January 2007.

MacKenzie, Donald. The Social Shaping of Technology. Bucks, England: Open University Press, 1999.

Mackie, J. Milton. From Cape Cod to Dixie and the Tropic. 2 volumes. New York: Putnam, 1864.

"Maddison Counting: A Long Passionate Affair with Numbers Has Finally Come to An End." The Economist. April 29, 2010. http://www.economist.com/research/articlesBySubject/displaystory.cfm?subjectid=348996&story_id=16004937, accessed 1 June 2010.

McCormick, Barnes, Conrad Newberry, and Eric Jumper, eds., Aerospace Engineering Education During the First Century of Flight. Reston, VA, American Institute of Aeronautics and Astronautics, 2004.

MacKenzie, Donald. Inventing Accuracy: A Historical Sociology of Nuclear Missile Guidance. Cambridge, Massachusetts,: MIT Press, 1990.

Morrison, Samuel Elliot. The Maritime History of Massachusetts, 1783-1860. Boston: Houhgton Mifflin, 1961.

Mystic Seaport Museum, http://www.mysticseaport.org/index.cfm?fuseaction=home.viewPage&page_id=B3E63C64-B3CA-45AE-A83D72C303A9C6BF.htm, accessed 9 September 2009.

NoseCone, www.nosecone.com/about.htm, accessed 3 June 2009.

http://www.nosecone.com/aepull.htm, accessed 31 December 2009.

"NOW THAT'S A BUMPER." March 2003. http://www.etrucker.com/apps/news/article.asp?id=34858.htm, accessed 3 June 2009.

"Parker Hannifin Joins Get Nitrogen Institute." 15 November 2006. http://www.truckinginfo.com/news/newsdetail.asp?news_id=57542&news_category_id=20.htm, accessed 11 January 2007.

Peebles, Curtis. The Forgotten X-Planes: Configuration Research Aircraft of the Supersonic Era, NASA Monographs in Aerospace History, forthcoming 2011.

Press Release, Lawrence Livermore National Laboratory. 2010 News Release. NR-10-02-08.

Overdrive. June 2005. http://www.etrucker.com/apps/news/article.asp?id=47566.htm, accessed 3 June 2009.

"Overdrive Foils Aerodynamic Wind Deflector." Overdrive. October 1974. vol. 14, no. 10, 88-91.

Russell, Phil. Navies in Transition: The Development of the Worlds Navies, the Technology, and the People Who Made It Happen. "Sir Charles A. Parsons, 1854-1931." http://www.btinternet.com/~philipr/Parsons.htm, accessed 29 September 2009.

Schenck, Paul. "New Focus on Air Drag." Trailer/Body Builders November 1975.

Sherman, Don, "Project Aerodynamic Van: Trying to Turn a Shoebox into a Slipper." Car and Driver April 1975.

Simon, Zoltan. The Double-Edged Sword: The Technological Sublime in American Novels, 1900-1940. Budapest: Akademia Kiado, 2003.

Smith, Merritt Roe. Harpers Ferry Armory and the New Technology: The Challenge of Change. Ithaca, N.Y.: Cornell University Press, 1977.

Smith, Merritt Roe, and Leo Marx, editors. Does Technology Drive History The Dilemma of Technological Determinism. Cambridge, MA: MIT Press, 1998.

Smith, William Laughton. The Pretensions of Thomas Jefferson to the Presidency Examined. (n.p., 1796) In Richard Hofstadter, Anti-intellectualism in American Life. New York: Alfred A. Knopf, 1963. 5th edition. Reprint, 1969.

Sorokin, Pitirim. "American Millionaires." Journal of Social Forces 3 1925.

"The New Mileage Misers." Brochure from HDT Heavy Duty Trucking. March 1975.

"The Shape of Trucks to Come." Trailer/Body Builders. [n.d.]

Tocqueville, Alexis de. Democracy in America. Translated by George Lawrence and edited by J. P. Meyer. Vol. 1 New York: Harper & Row. Reprint by Anchor Books, 1969.

Bibliography

Toon, John. "Flying Low-Drag Trucks: Aerodynamic concepts and controls for aircraft will cut fuel use and improve control in trucks." Georgia Tech Research Horizons, 16 February 2001. http://gtresearchnews.gatech.edu/reshor/rh-win01/trucks.htm, accessed 3 June 2009.

Reed, R. Dale with Darlene Lister. Wingless Flight: The Lifting Body Story. Lexington: the University of Kentucky Press, 2002.

Taylor, Frederick W. The Principles of Scientific Management. New York, N.Y.: W. W. Norton, 1967.

Thompson, Milton O., with a background section by J. D. Hunley. Flight Research: Problems Encountered and What They Should Teach Us. NASA SP 2000-4522, 2000.

Transport Energy Best Practices: Smoothing the Flow at TNT Express and Somerfield using Truck Aerodynamics. [Crown? n.p., n.d.].

Transport Energy Best Practices: The Streamlined Guide to Truck Aerodynamic Styling, Department of Transport. Crown, UK, 2004.

"Utility Trailer side skirts receive SmartWay verification," FleetOwner, Feb. 18, 2010, http://fleetowner.com/equipment/news/utility-trailer-side-skirts-0218/, accessed 2 June 2010.

Veblen, Thorstein. The Theory of the Leisure Class: An Economic Study of Institutions. With a Foreword by Stuart Chase. New York: The Modern Library, 1899. Reprint, 1934.

"Volvo Displays Aerodynamic Devices For Improved Fuel Economy." 15 November 2006. http://trailer-bodybuilders.com/news/volvo_fuel_economy/htm, accessed 3 June 2009.

"Wal-Mart May Save $300 Million with Fleet Efficiency," Clean Fleet Report, August 4, 2009, http://www.cleanfleetreport.com/fleets/wal-mart-to-save-300-million-with-hybrids/, accessed 2 June 2010.

Winning the Oil End Game. Technology Annex, chapter 6. "Class 8 Heavy Trucks." 6: http://www.oilendgame.org, accessed 18 June 2009.

Wright, Mike. "The Disney-Von Braun Collaboration and Its Influence on Space Exploration." http://history.msfc.nasa.gov/vonbraun/disney_article.html accessed 28 April 2011.

Wyllie, Irving G. The Self-Made Man In America: The Myth of Rags to Riches. New York: The Free Press, 1966.

Unpublished Sources

Altrichter, Kirk, Vice President, Maintenance, Gordon Trucking, International, telephone interview with the author, 19 July 2010.

Bowers, Albion interview with the author, NASA Dryden Flight Research Center, Edwards, CA, 4 December 2009.

Correspondence from Larry Cagan, Stanford Research Institute, Menlo Park, CA, to Bud Hartman, NASA Headquarters, Washington, D.C., 24 December 1980. In the private collection of Edwin J. Saltzman.

MacDonald, Alexander. "The Long Space Age: Essays on the Economic History of Space Exploration from Galileo to Gagarin," D.Phil. diss., Baliol College, Oxford University, 2010.

Moes, Tim, Tony Whitmore, Ken Illif, and Ed Saltzman. Memorandum to Code RA, etc. "Base drag Reduction using a Ground Research Vehicle." 12 October 2000.

Memorandum to Director regarding "Request for approval of research and development project—study the increase of efficiency of ground vehicles, from E. J. Saltzman and R. R. Meyer, 22 November 1972," in the Dryden Historical Collection.

Notes, "Low Drag Truck Design Tested by NASA," NASA Ames Research Center (n.d.).

Saltzman, Edwin J. "A Summary of NASA Dryden's Aerodynamic Truck Research" to be presented at the 1982 SAE Truck and Bus Meeting and Exposition, Indianapolis, IN, 8-11 November 1982.

_____. Comments attached to various NASA Technology notes on file in the Dryden Historical Collection.

_____. Interview with the author, Dryden Flight Research Center, Edwards, CA, 5 September 2003. Transcript in the Dryden Historical Collection.

_____. Memorandum: "Request for Project Approval (RFPA) of a Truck Aerodynamic Study." To Director of Research at NASA Flight Research Center, Edwards, CA, February 20, 1974.

_____. to Gelzer, notes on manuscript draft.

Smith, Derek. PACCAR, Inc., electronic mail correspondence with author, 16 September 2003. Transcript in the Dryden Historical Collection.

Smith, Stephanie, Ph.D. Auburn University History Department, interview with the author, April 2002.

Index

Aerospan - 39, 40

AeroVironment - 38, 39, 58

Airshield - 17, 22, 39, 40, 53, 56, 57

AirTabs - 57

Altrichter, Kirk - 59

Base drag - 2, 4, 5, 11, 30, 39, 52, 53, 57, 61, 62, 63, 65, 66

Bat Truck - 25, 26, 27, 29, 30, 31

Boattail - 14, 32, 33, 34, 35, 50, 51, 53, 58, 59, 60, 63

Bridge formula - 20

Cattle - 43, 44, 45, 46, 47, 61

CFI - 41, 42

Coast-down - 12, 13, 18, 21, 22, 27, 28, 30, 39, 62

Drag Bucket - 51, 52, 65, 66

FitzGerald Nose Cone (also NoseCone) - 17, 18

Ford - 6, 11, 13, 22, 57

Forebody drag - 52, 53, 62, 63, 65, 66

FRC - 1, 7, 8, 12, 15, 17, 18, 19, 21, 22, 23, 25, 27, 28, 30, 31, 32, 33, 34, 37, 38, 39, 46, 49, 52, 53, 55, 57, 58, 61, 67, 70

Freightliner - 17, 30, 40

Freight Wing - 59, 60

General Motors - 22, 30, 42

Gordon Trucking, Inc. - 59

Hoerner - 6, 12, 13, 63

Horn, Floyd - 43

Horton, Vic - 11

Illif, Ken - 62

Kansas - 22, 33, 34, 35, 46

Kenworth - 26, 41, 42, 50, 56, 57, 58, 60, 67

Lifting Bodies - 1, 8, 23, 51, 61

Lissaman, P. - 5, 38, 39

Low pressure - 5

Mack - 57, 58, 69

Meyer, Robert - 11, 12, 52, 63

Moes, Timothy - 62

Montoya, Lawrence - 18

Moses, Robert - vii, viii

Muirhead, Vincent - 7, 34

NACA - 1, 3, 4, 23, 26, 37, 47

Peterbilt - 50, 56

Roadrunner - 62, 63, 64, 65, 66

Rocket - vii, 1, 3, 4, 8, 9, 12, 61, 70

Rudkin Wylie - 17, 20, 22, 40, 53, 57

Ryder - 41

Saltzman, Ed - 4, 5, 7, 11, 43, 62

Schenck, Paul - 20, 39

Sherman, Don - 37

Shoebox - 11, 12, 13, 14, 15, 17, 18, 19, 22, 27, 28, 29, 30, 32, 33, 34, 37, 42, 50, 51, 52, 53, 59, 60, 62, 63

South Base - 11, 13, 17, 21, 27

Steers, Louis - 17, 18

Thompson, Milt - 11

Tufting - 27, 29, 32

US DoT - 49

Volvo - 57

Vortex - 17, 40, 57, 58

Vortices - 5, 32, 57, 58

Whitmore, Stephen (Tony) - 61, 62

X-15 - 1, 3, 4, 8, 9, 12, 23, 37, 53

The NASA History Series

REFERENCE WORKS, NASA SP-4000:

Grimwood, James M. *Project Mercury: A Chronology*. NASA SP-4001, 1963.

Grimwood, James M., and Barton C. Hacker, with Peter J. Vorzimmer. *Project Gemini Technology and Operations: A Chronology*. NASA SP-4002, 1969.

Link, Mae Mills. *Space Medicine in Project Mercury*. NASA SP-4003, 1965.

Astronautics and Aeronautics, 1963: Chronology of Science, Technology, and Policy. NASA SP-4004, 1964.

Astronautics and Aeronautics, 1964: Chronology of Science, Technology, and Policy. NASA SP-4005, 1965.

Astronautics and Aeronautics, 1965: Chronology of Science, Technology, and Policy. NASA SP-4006, 1966.

Astronautics and Aeronautics, 1966: Chronology of Science, Technology, and Policy. NASA SP-4007, 1967.

Astronautics and Aeronautics, 1967: Chronology of Science, Technology, and Policy. NASA SP-4008, 1968.

Ertel, Ivan D., and Mary Louise Morse. *The Apollo Spacecraft: A Chronology, Volume I, Through November 7, 1962*. NASA SP-4009, 1969.

Morse, Mary Louise, and Jean Kernahan Bays. *The Apollo Spacecraft: A Chronology, Volume II, November 8, 1962–September 30, 1964*. NASA SP-4009, 1973.

Brooks, Courtney G., and Ivan D. Ertel. *The Apollo Spacecraft: A Chronology, Volume III, October 1, 1964–January 20, 1966*. NASA SP-4009, 1973.

Ertel, Ivan D., and Roland W. Newkirk, with Courtney G. Brooks. *The Apollo Spacecraft: A Chronology, Volume IV, January 21, 1966–July 13, 1974*. NASA SP-4009, 1978.

Astronautics and Aeronautics, 1968: Chronology of Science, Technology, and Policy. NASA SP-4010, 1969.

Newkirk, Roland W., and Ivan D. Ertel, with Courtney G. Brooks. *Skylab: A Chronology*. NASA SP-4011, 1977.

Van Nimmen, Jane, and Leonard C. Bruno, with Robert L. Rosholt. *NASA Historical Data Book, Vol. I: NASA Resources, 1958–1968*. NASA SP-4012, 1976, rep. ed. 1988.

Ezell, Linda Neuman. *NASA Historical Data Book, Vol. II: Programs and Projects, 1958–1968*. NASA SP-4012, 1988.

Ezell, Linda Neuman. *NASA Historical Data Book, Vol. III: Programs and Projects, 1969–1978*. NASA SP-4012, 1988.

Gawdiak, Ihor, with Helen Fedor. *NASA Historical Data Book, Vol. IV: NASA Resources, 1969–1978.* NASA SP-4012, 1994.

Rumerman, Judy A. *NASA Historical Data Book, Vol. V: NASA Launch Systems, Space Transportation, Human Spaceflight, and Space Science, 1979–1988.* NASA SP-4012, 1999.

Rumerman, Judy A. *NASA Historical Data Book, Vol. VI: NASA Space Applications, Aeronautics and Space Research and Technology, Tracking and Data Acquisition/Support Operations, Commercial Programs, and Resources, 1979–1988.* NASA SP-4012, 1999.

Rumerman, Judy A. *NASA Historical Data Book, Vol. VII: NASA Launch Systems, Space Transportation, Human Spaceflight, and Space Science, 1989–1998.* NASA SP-2009-4012.

No SP-4013.

Astronautics and Aeronautics, 1969: Chronology of Science, Technology, and Policy. NASA SP-4014, 1970.

Astronautics and Aeronautics, 1970: Chronology of Science, Technology, and Policy. NASA SP-4015, 1972.

Astronautics and Aeronautics, 1971: Chronology of Science, Technology, and Policy. NASA SP-4016, 1972.

Astronautics and Aeronautics, 1972: Chronology of Science, Technology, and Policy. NASA SP-4017, 1974.

Astronautics and Aeronautics, 1973: Chronology of Science, Technology, and Policy. NASA SP-4018, 1975.

Astronautics and Aeronautics, 1974: Chronology of Science, Technology, and Policy. NASA SP-4019, 1977.

Astronautics and Aeronautics, 1975: Chronology of Science, Technology, and Policy. NASA SP-4020, 1979.

Astronautics and Aeronautics, 1976: Chronology of Science, Technology, and Policy. NASA SP-4021, 1984.

Astronautics and Aeronautics, 1977: Chronology of Science, Technology, and Policy. NASA SP-4022, 1986.

Astronautics and Aeronautics, 1978: Chronology of Science, Technology, and Policy. NASA SP-4023, 1986.

Astronautics and Aeronautics, 1979–1984: Chronology of Science, Technology, and Policy. NASA SP-4024, 1988.

Astronautics and Aeronautics, 1985: Chronology of Science, Technology, and Policy. NASA SP-4025, 1990.

Noordung, Hermann. *The Problem of Space Travel: The Rocket Motor.* Edited by Ernst Stuhlinger and J.D. Hunley, with Jennifer Garland. NASA SP-4026, 1995.

Astronautics and Aeronautics, 1986–1990: A Chronology. NASA SP-4027, 1997.

Astronautics and Aeronautics, 1991–1995: A Chronology. NASA SP-2000-4028, 2000.

Orloff, Richard W. *Apollo by the Numbers: A Statistical Reference*. NASA SP-2000-4029, 2000.

Lewis, Marieke and Swanson, Ryan. *Aeronautics and Astronautics: A Chronology, 1996-2000*. NASA SP-2009-4030, 2009.

Ivey, William Noel and Swanson, Ryan. *Aeronautics and Astronautics: A Chronology, 2001-2005*. NASA SP-2010-4031, 2010.

MANAGEMENT HISTORIES, NASA SP-4100:

Rosholt, Robert L. *An Administrative History of NASA, 1958–1963*. NASA SP-4101, 1966.

Levine, Arnold S. *Managing NASA in the Apollo Era*. NASA SP-4102, 1982.

Roland, Alex. *Model Research: The National Advisory Committee for Aeronautics, 1915–1958*. NASA SP-4103, 1985.

Fries, Sylvia D. *NASA Engineers and the Age of Apollo*. NASA SP-4104, 1992.

Glennan, T. Keith. *The Birth of NASA: The Diary of T. Keith Glennan*. Edited by J.D. Hunley. NASA SP-4105, 1993.

Seamans, Robert C. *Aiming at Targets: The Autobiography of Robert C. Seamans*. NASA SP-4106, 1996.

Garber, Stephen J., editor. *Looking Backward, Looking Forward: Forty Years of Human Spaceflight Symposium*. NASA SP-2002-4107, 2002.

Mallick, Donald L. with Peter W. Merlin. *The Smell of Kerosene: A Test Pilot's Odyssey*. NASA SP-4108, 2003.

Iliff, Kenneth W. and Curtis L. Peebles. *From Runway to Orbit: Reflections of a NASA Engineer*. NASA SP-2004-4109, 2004.

Chertok, Boris. *Rockets and People, Volume 1*. NASA SP-2005-4110, 2005.

Chertok, Boris. *Rockets and People: Creating a Rocket Industry, Volume II*. NASA SP-2006-4110, 2006.

Chertok, Boris. *Rockets and People: Hot Days of the Cold War, Volume III*. NASA SP-2009-4110, 2009.

Laufer, Alexander, Todd Post, and Edward Hoffman. *Shared Voyage: Learning and Unlearning from Remarkable Projects*. NASA SP-2005-4111, 2005.

Dawson, Virginia P., and Mark D. Bowles. *Realizing the Dream of Flight: Biographical Essays in Honor of the Centennial of Flight, 1903–2003*. NASA SP-2005-4112, 2005.

Mudgway, Douglas J. William H. Pickering: *America's Deep Space Pioneer*. NASA SP-2008-4113.

PROJECT HISTORIES, NASA SP-4200:

Swenson, Loyd S., Jr., James M. Grimwood, and Charles C. Alexander. *This New Ocean: A History of Project Mercury.* NASA SP-4201, 1966; reprinted 1999.

Green, Constance McLaughlin, and Milton Lomask. *Vanguard: A History.* NASA SP-4202, 1970; rep. ed. Smithsonian Institution Press, 1971.

Hacker, Barton C., and James M. Grimwood. *On Shoulders of Titans: A History of Project Gemini.* NASA SP-4203, 1977, reprinted 2002.

Benson, Charles D., and William Barnaby Faherty. *Moonport: A History of Apollo Launch Facilities and Operations.* NASA SP-4204, 1978.

Brooks, Courtney G., James M. Grimwood, and Loyd S. Swenson, Jr. *Chariots for Apollo: A History of Manned Lunar Spacecraft.* NASA SP-4205, 1979.

Bilstein, Roger E. *Stages to Saturn: A Technological History of the Apollo/Saturn Launch Vehicles.* NASA SP-4206, 1980 and 1996.

No SP-4207.

Compton, W. David, and Charles D. Benson. *Living and Working in Space: A History of Skylab.* NASA SP-4208, 1983.

Ezell, Edward Clinton, and Linda Neuman Ezell. *The Partnership: A History of the Apollo-Soyuz Test Project.* NASA SP-4209, 1978.

Hall, R. Cargill. *Lunar Impact: A History of Project Ranger.* NASA SP-4210, 1977.

Newell, Homer E. *Beyond the Atmosphere: Early Years of Space Science.* NASA SP-4211, 1980.

Ezell, Edward Clinton, and Linda Neuman Ezell. *On Mars: Exploration of the Red Planet, 1958–1978.* NASA SP-4212, 1984.

Pitts, John A. *The Human Factor: Biomedicine in the Manned Space Program to 1980.* NASA SP-4213, 1985.

Compton, W. David. *Where No Man Has Gone Before: A History of Apollo Lunar Exploration Missions.* NASA SP-4214, 1989.

Naugle, John E. *First Among Equals: The Selection of NASA Space Science Experiments.* NASA SP-4215, 1991.

Wallace, Lane E. *Airborne Trailblazer: Two Decades with NASA Langley's 737 Flying Laboratory.* NASA SP-4216, 1994.

Butrica, Andrew J., ed. *Beyond the Ionosphere: Fifty Years of Satellite Communications.* NASA SP-4217, 1997.

Butrica, Andrew J. *To See the Unseen: A History of Planetary Radar Astronomy.* NASA SP-4218, 1996.

Mack, Pamela E., ed. *From Engineering Science to Big Science: The NACA and NASA Collier Trophy Research Project Winners*. NASA SP-4219, 1998.

Reed, R. Dale. *Wingless Flight: The Lifting Body Story*. NASA SP-4220, 1998.

Heppenheimer, T. A. *The Space Shuttle Decision: NASA's Search for a Reusable Space Vehicle*. NASA SP-4221, 1999.

Hunley, J. D., ed. *Toward Mach 2: The Douglas D-558 Program*. NASA SP-4222, 1999.

Swanson, Glen E., ed. *"Before This Decade is Out . . ." Personal Reflections on the Apollo Program*. NASA SP-4223, 1999.

Tomayko, James E. *Computers Take Flight: A History of NASA's Pioneering Digital Fly-By-Wire Project*. NASA SP-4224, 2000.

Morgan, Clay. *Shuttle-Mir: The United States and Russia Share History's Highest Stage*. NASA SP-2001-4225.

Leary, William M. *"We Freeze to Please:" A History of NASA's Icing Research Tunnel and the Quest for Safety*. NASA SP-2002-4226, 2002.

Mudgway, Douglas J. *Uplink-Downlink: A History of the Deep Space Network, 1957–1997*. NASA SP-2001-4227.

No SP-4228 or SP-4229.

Dawson, Virginia P., and Mark D. Bowles. *Taming Liquid Hydrogen: The Centaur Upper Stage Rocket, 1958–2002*. NASA SP-2004-4230.

Meltzer, Michael. *Mission to Jupiter: A History of the Galileo Project*. NASA SP-2007-4231.

Heppenheimer, T. A. *Facing the Heat Barrier: A History of Hypersonics*. NASA SP-2007-4232.

Tsiao, Sunny. *"Read You Loud and Clear!" The Story of NASA's Spaceflight Tracking and Data Network*. NASA SP-2007-4233.

CENTER HISTORIES, NASA SP-4300:

Rosenthal, Alfred. *Venture into Space: Early Years of Goddard Space Flight Center*. NASA SP-4301, 1985.

Hartman, Edwin, P. *Adventures in Research: A History of Ames Research Center, 1940–1965*. NASA SP-4302, 1970.

Hallion, Richard P. *On the Frontier: Flight Research at Dryden, 1946–1981*. NASA SP-4303, 1984.

Muenger, Elizabeth A. *Searching the Horizon: A History of Ames Research Center, 1940–1976*. NASA SP-4304, 1985.

Hansen, James R. *Engineer in Charge: A History of the Langley Aeronautical Laboratory, 1917–1958.* NASA SP-4305, 1987.

Dawson, Virginia P. *Engines and Innovation: Lewis Laboratory and American Propulsion Technology.* NASA SP-4306, 1991.

Dethloff, Henry C. *"Suddenly Tomorrow Came . . .": A History of the Johnson Space Center, 1957–1990.* NASA SP-4307, 1993.

Hansen, James R. *Spaceflight Revolution: NASA Langley Research Center from Sputnik to Apollo.* NASA SP-4308, 1995.

Wallace, Lane E. *Flights of Discovery: Fifty Years of Flight Research at Dryden flight Research Center.* NASA SP-4309, 2006.

Herring, Mack R. *Way Station to Space: A History of the John C. Stennis Space Center.* NASA SP-4310, 1997.

Wallace, Harold D., Jr. *Wallops Station and the Creation of an American Space Program.* NASA SP-4311, 1997.

Wallace, Lane E. *Dreams, Hopes, Realities. NASA's Goddard Space Flight Center: The First Forty Years.* NASA SP-4312, 1999.

Dunar, Andrew J., and Stephen P. Waring. *Power to Explore: A History of Marshall Space Flight Center, 1960–1990.* NASA SP-4313, 1999.

Bugos, Glenn E. *Atmosphere of Freedom: Sixty Years at the NASA Ames Research Center.* NASA SP-2000-4314, 2000.

NO SP-4315.

Schultz, James. *Crafting Flight: Aircraft Pioneers and the Contributions of the Men and Women of NASA Langley Research Center.* NASA SP-2003-4316, 2003.

Bowles, Mark D. *Science in Flux: NASA's Nuclear Program at Plum Brook Station, 1955–2005.* NASA SP-2006-4317.

Wallace, Lane E. *Flights of Discovery: Sixty Years of Flight Research at Dryden Flight Research Center.* NASA SP-2006-4318, 2006. Revised version of SP-4309.

Arrighi, Robert S. *Revolutionary Atmosphere: The Story of the Altitude Wind Tunnel and the Space Power Chambers.* NASA SP-2010-4319.

GENERAL HISTORIES, NASA SP-4400:

Corliss, William R. *NASA Sounding Rockets, 1958–1968: A Historical Summary.* NASA SP-4401, 1971.

Wells, Helen T., Susan H. Whiteley, and Carrie Karegeannes. *Origins of NASA Names.* NASA SP-4402, 1976.

Anderson, Frank W., Jr. *Orders of Magnitude: A History of NACA and NASA, 1915–1980*. NASA SP-4403, 1981.

Sloop, John L. *Liquid Hydrogen as a Propulsion Fuel, 1945–1959*. NASA SP-4404, 1978.

Roland, Alex. *A Spacefaring People: Perspectives on Early Spaceflight*. NASA SP-4405, 1985.

Bilstein, Roger E. *Orders of Magnitude: A History of the NACA and NASA, 1915–1990*. NASA SP-4406, 1989.

Logsdon, John M., ed., with Linda J. Lear, Jannelle Warren Findley, Ray A. Williamson, and Dwayne A. Day. *Exploring the Unknown: Selected Documents in the History of the U.S. Civil Space Program, Volume I, Organizing for Exploration*. NASA SP-4407, 1995.

Logsdon, John M., ed, with Dwayne A. Day, and Roger D. Launius. *Exploring the Unknown: Selected Documents in the History of the U.S. Civil Space Program, Volume II, External Relationships*. NASA SP-4407, 1996.

Logsdon, John M., ed., with Roger D. Launius, David H. Onkst, and Stephen J. Garber. *Exploring the Unknown: Selected Documents in the History of the U.S. Civil Space Program, Volume III, Using Space*. NASA SP-4407,1998.

Logsdon, John M., ed., with Ray A. Williamson, Roger D. Launius, Russell J. Acker, Stephen J. Garber, and Jonathan L. Friedman. *Exploring the Unknown: Selected Documents in the History of the U.S. Civil Space Program, Volume IV, Accessing Space*. NASA SP-4407, 1999.

Logsdon, John M., ed., with Amy Paige Snyder, Roger D. Launius, Stephen J. Garber, and Regan Anne Newport. *Exploring the Unknown: Selected Documents in the History of the U.S. Civil Space Program, Volume V, Exploring the Cosmos*. NASA SP-4407, 2001.

Logsdon, John M., ed., with Stephen J. Garber, Roger D. Launius, and Ray A. Williamson. *Exploring the Unknown: Selected Documents in the History of the U.S. Civil Space Program, Volume VI: Space and Earth Science*. NASA SP-2004-4407, 2004.

Logsdon, John M., ed., with Roger D. Launius. *Exploring the Unknown: Selected Documents in the History of the U.S. Civil Space Program, Volume VII: Human Spaceflight: Projects Mercury, Gemini, and Apollo*. NASA SP-2008-4407, 2008.

Siddiqi, Asif A., *Challenge to Apollo: The Soviet Union and the Space Race, 1945–1974*. NASA SP-2000-4408, 2000.

Hansen, James R., ed. *The Wind and Beyond: Journey into the History of Aerodynamics in America, Volume 1, The Ascent of the Airplane*. NASA SP-2003-4409, 2003.

Hansen, James R., ed. *The Wind and Beyond: Journey into the History of Aerodynamics in America, Volume 2, Reinventing the Airplane*. NASA SP-2007-4409, 2007.

Hogan, Thor. *Mars Wars: The Rise and Fall of the Space Exploration Initiative*. NASA SP-2007-4410, 2007.

The NASA History Series

MONOGRAPHS IN AEROSPACE HISTORY (SP-4500 SERIES):

Launius, Roger D., and Aaron K. Gillette, compilers. *Toward a History of the Space Shuttle: An Annotated Bibliography.* Monograph in Aerospace History, No. 1, 1992.

Launius, Roger D., and J. D. Hunley, compilers. *An Annotated Bibliography of the Apollo Program.* Monograph in Aerospace History No. 2, 1994.

Launius, Roger D. *Apollo: A Retrospective Analysis.* Monograph in Aerospace History, No. 3, 1994.

Hansen, James R. *Enchanted Rendezvous: John C. Houbolt and the Genesis of the Lunar-Orbit Rendezvous Concept.* Monograph in Aerospace History, No. 4, 1995.

Gorn, Michael H. *Hugh L. Dryden's Career in Aviation and Space.* Monograph in Aerospace History, No. 5, 1996.

Powers, Sheryll Goecke. *Women in Flight Research at NASA Dryden Flight Research Center from 1946 to 1995.* Monograph in Aerospace History, No. 6, 1997.

Portree, David S. F., and Robert C. Trevino. *Walking to Olympus: An EVA Chronology.* Monograph in Aerospace History, No. 7, 1997.

Logsdon, John M., moderator. *Legislative Origins of the National Aeronautics and Space Act of 1958: Proceedings of an Oral History Workshop.* Monograph in Aerospace History, No. 8, 1998.

Rumerman, Judy A., compiler. *U.S. Human Spaceflight, A Record of Achievement 1961–1998.* Monograph in Aerospace History, No. 9, 1998.

Portree, David S. F. *NASA's Origins and the Dawn of the Space Age.* Monograph in Aerospace History, No. 10, 1998.

Logsdon, John M. *Together in Orbit: The Origins of International Cooperation in the Space Station.* Monograph in Aerospace History, No. 11, 1998.

Phillips, W. Hewitt. *Journey in Aeronautical Research: A Career at NASA Langley Research Center.* Monograph in Aerospace History, No. 12, 1998.

Braslow, Albert L. *A History of Suction-Type Laminar-Flow Control with Emphasis on Flight Research.* Monograph in Aerospace History, No. 13, 1999.

Logsdon, John M., moderator. *Managing the Moon Program: Lessons Learned From Apollo.* Monograph in Aerospace History, No. 14, 1999.

Perminov, V. G. *The Difficult Road to Mars: A Brief History of Mars Exploration in the Soviet Union.* Monograph in Aerospace History, No. 15, 1999.

Tucker, Tom. *Touchdown: The Development of Propulsion Controlled Aircraft at NASA Dryden.* Monograph in Aerospace History, No. 16, 1999.

Maisel, Martin, Demo J.Giulanetti, and Daniel C. Dugan. *The History of the XV-15 Tilt Rotor Research Aircraft: From Concept to Flight.* Monograph in Aerospace History, No. 17, 2000. NASA SP-2000-4517.

Jenkins, Dennis R. *Hypersonics Before the Shuttle: A Concise History of the X-15 Research Airplane.* Monograph in Aerospace History, No. 18, 2000. NASA SP-2000-4518.

Chambers, Joseph R. *Partners in Freedom: Contributions of the Langley Research Center to U.S. Military Aircraft of the 1990s.* Monograph in Aerospace History, No. 19, 2000. NASA SP-2000-4519.

Waltman, Gene L. *Black Magic and Gremlins: Analog Flight Simulations at NASA's Flight Research Center.* Monograph in Aerospace History, No. 20, 2000. NASA SP-2000-4520.
Portree, David S. F. *Humans to Mars: Fifty Years of Mission Planning, 1950–2000.* Monograph in Aerospace History, No. 21, 2001. NASA SP-2001-4521.

Thompson, Milton O., with J. D. Hunley. *Flight Research: Problems Encountered and What they Should Teach Us.* Monograph in Aerospace History, No. 22, 2001. NASA SP-2001-4522.

Tucker, Tom. *The Eclipse Project.* Monograph in Aerospace History, No. 23, 2001. NASA SP-2001-4523.

Siddiqi, Asif A. *Deep Space Chronicle: A Chronology of Deep Space and Planetary Probes 1958–2000.* Monograph in Aerospace History, No. 24, 2002. NASA SP-2002-4524.

Merlin, Peter W. *Mach 3+: NASA/USAF YF-12 Flight Research, 1969–1979.* Monograph in Aerospace History, No. 25, 2001. NASA SP-2001-4525.

Anderson, Seth B. *Memoirs of an Aeronautical Engineer: Flight Tests at Ames Research Center: 1940–1970.* Monograph in Aerospace History, No. 26, 2002. NASA SP-2002-4526.

Renstrom, Arthur G. *Wilbur and Orville Wright: A Bibliography Commemorating the One-Hundredth Anniversary of the First Powered Flight on December 17, 1903.* Monograph in Aerospace History, No. 27, 2002. NASA SP-2002-4527.

No monograph 28.

Chambers, Joseph R. *Concept to Reality: Contributions of the NASA Langley Research Center to U.S. Civil Aircraft of the 1990s.* Monograph in Aerospace History, No. 29, 2003. SP-2003-4529.

Peebles, Curtis, editor. *The Spoken Word: Recollections of Dryden History, The Early Years.* Monograph in Aerospace History, No. 30, 2003. SP-2003-4530.

Jenkins, Dennis R., Tony Landis, and Jay Miller. *American X-Vehicles: An Inventory- X-1 to X-50.* Monograph in Aerospace History, No. 31, 2003. SP-2003-4531.

Renstrom, Arthur G. *Wilbur and Orville Wright: A Chronology Commemorating the One-Hundredth Anniversary of the First Powered Flight on December 17, 1903.* Monograph in Aerospace History, No. 32, 2003. NASA SP-2003-4532.

Bowles, Mark D., and Robert S. Arrighi. *NASA's Nuclear Frontier: The Plum Brook Research Reactor.* Monograph in Aerospace History, No. 33, 2004. (SP-2004-4533).

Wallace, Lane and Christian Gelzer. *Nose Up: High Angle-of-Attack and Thrust Vectoring Research at NASA Dryden, 1979-2001.* Monograph in Aerospace History No. 34, 2009. NASA SP-2009-4534.

Matranga, Gene J., C. Wayne Ottinger, Calvin R. Jarvis, and D. Christian Gelzer. *Unconventional, Contrary, and Ugly: The Lunar Landing Research Vehicle.* Monograph in Aerospace History, No. 35, 2006. NASA SP-2004-4535.

McCurdy, Howard E. *Low Cost Innovation in Spaceflight: The History of the Near Earth Asteroid Rendezvous (NEAR) Mission.* Monograph in Aerospace History, No. 36, 2005. NASA SP-2005-4536.

Seamans, Robert C., Jr. *Project Apollo: The Tough Decisions.* Monograph in Aerospace History, No. 37, 2005. NASA SP-2005-4537.

Lambright, W. Henry. *NASA and the Environment: The Case of Ozone Depletion.* Monograph in Aerospace History, No. 38, 2005. NASA SP-2005-4538.

Chambers, Joseph R. *Innovation in Flight: Research of the NASA Langley Research Center on Revolutionary Advanced Concepts for Aeronautics.* Monograph in Aerospace History, No. 39, 2005. NASA SP-2005-4539.

Phillips, W. Hewitt. *Journey Into Space Research: Continuation of a Career at NASA Langley Research Center.* Monograph in Aerospace History, No. 40, 2005. NASA SP-2005-4540.

Rumerman, Judy A., Chris Gamble, and Gabriel Okolski, compilers. *U.S. Human Spaceflight: A Record of Achievement, 1961–2006.* Monograph in Aerospace History No. 41, 2007. NASA SP-2007-4541.

Dick, Steven J.; Garber, Stephen J.; and Odom, Jane H. *Research in NASA History.* Monograph in Aerospace History No. 43, 2009. NASA SP-2009-4543.

Merlin, Peter W., Ikhana: *Unmanned Aircraft System Western States Fire Missions.* Monograph in Aerospace History #44. NASA SP-2009-4544.

Fisher, Steven C. and Rahman, Shamim A. *Remembering the Giants: Apollo Rocket Propulsion Development.* Monograph in Aerospace History #45. NASA SP-2009-4545.

ELECTRONIC MEDIA (SP-4600 SERIES):

Remembering Apollo 11: The 30th Anniversary Data Archive CD-ROM. NASA SP-4601, 1999.

Remembering Apollo 11: The 35th Anniversary Data Archive CD-ROM. NASA SP-2004-4601, 2004. This is an update of the 1999 edition.

The Mission Transcript Collection: U.S. Human Spaceflight Missions from Mercury Redstone 3 to Apollo 17. SP-2000-4602, 2001.

Shuttle-Mir: the United States and Russia Share History's Highest Stage. NASA SP-2001-4603, 2002.

U.S. Centennial of Flight Commission presents Born of Dreams ~ Inspired by Freedom. NASA SP-2004-4604, 2004.

Of Ashes and Atoms: A Documentary on the NASA Plum Brook Reactor Facility. NASA SP-2005-4605.

Taming Liquid Hydrogen: The Centaur Upper Stage Rocket Interactive CD-ROM. NASA SP-2004-4606, 2004.

Fueling Space Exploration: The History of NASA's Rocket Engine Test Facility DVD. NASA SP-2005-4607.

Altitude Wind Tunnel at NASA Glenn Research Center: An Interactive History CD-ROM. NASA SP-2008-4608.

A Tunnel Through Time: The History of NASA's Altitude Wind Tunnel. NASA SP-2010-4609.

CONFERENCE PROCEEDINGS (SP-4700 SERIES):

Dick, Steven J., and Keith Cowing, ed. *Risk and Exploration: Earth, Sea and the Stars.* NASA SP-2005-4701.

Dick, Steven J., and Roger D. Launius. *Critical Issues in the History of Spaceflight.* NASA SP-2006-4702.

Dick, Steven J., ed. *Remembering the Space Age: Proceedings of the 50th Anniversary Conference.* NASA SP-2008-4703.

Dick, Steven J., editor, *NASA's First 50 Years: Historical Perspectives.* NASA SP-2010-4704.

SOCIETAL IMPACT (SP-4800 SERIES):

Dick, Steven J., and Roger D. Launius. *Societal Impact of Spaceflight.* NASA SP-2007-4801.

Dick, Steven J., and Lupisella, Mark L. *Cosmos and Culture: Cultural Evolution in a Cosmic Context.* NASA SP-2009-4802.

www.ingramcontent.com/pod-product-compliance
Lightning Source LLC
Chambersburg PA
CBHW080508110426
42742CB00017B/3039